GENE STRUCTURE AND TRANSCRIPTION

Trevor Beebee and Julian Burke

Biochemistry Group, School of Biological Sciences, Biology Building, University of Sussex, Falmer, Brighton BN1 9QG, UK

OXFORD · WASHINGTON DC

IRL Press
Eynsham
Oxford
England

© IRL Press Limited 1988

First published 1988

All rights reserved by the publisher. No part of this book may be reproduced or transmitted in any form by any means, electronic or mechanical, including photocopying, recording or any information storage and retrieval system, without permission in writing from the publisher.

British Library Cataloguing in Publication Data

Beebee, T. (Trevor)
    Gene structure and transcription.
    1. Organisms. Genes
    I. Title   II. Burke, J. (Julian)
    III. Series
    575.1′2

Library of Congress Cataloging in Publication Data

Beebee, Trevor J. C. (Trevor John Clark)
    Gene structure and transcription.

    (In focus)
    Includes bibliographies and index.
    1. genetic transcription. I. Burke, Julian.
II. Title.
QH450.2.B44   1988   574.87′328   88-13258
ISBN 1 85221 014 1 (softbound)

Typeset by Infotype and printed by Information Printing Ltd, Oxford, England.

Preface

Molecular biology and, in particular, the study of gene expression, is still firmly in the exponential growth phase despite the fact that more than 40 years have passed since the demonstration that DNA carries genetic information. Achievements have been dramatic; the techniques of DNA cloning and sequencing continue to produce data in quantities inconceivable 20 years ago, and uninterpretable without concurrent developments in computer technology. We have an increasingly clear idea of what bacterial and eukaryotic genomes look like; several viruses have had their genomes completely sequenced and that of the common gut bacterium *Escherichia coli* will soon follow. Now plans are afoot to escalate sequencing efforts by further orders of magnitude, with the entire human genome—the daunting task of some three billion base pairs—set squarely in our sights.

Yet many of the most haunting problems of biology remain elusive. We still do not know, in any real sense of the word, how multicellular organisms develop from a fertilized egg. Despite recent dramatic discoveries in the field of oncogenes, our understanding of carcinogenesis and other degenerative diseases remains at a rudimentary, albeit increasingly exciting level. Molecular biology is only beginning to fulfil its great promise in areas such as evolution and speciation, where it threatens to replace at least some areas of classical biology altogether.

Detailed knowledge of gene structure and the transcription process will be central to solving the major outstanding problems of biology indicated above. In such a busy research scene, where new journals and books are emerging almost monthly, no text can expect to stay on top for long. One way around this problem has been attempted in this series: relatively short texts, concentrating on what are judged to be the most significant areas of research, that can be updated quickly and regularly. Indeed, it was daunting to realize how often the text needed adjustment to take account of papers being published even during the preparation of this manuscript!

To accomplish our aims, we have set out to emphasize similarities and differences between prokaryotes (relatively simple and well studied) and eukaryotes (much more complex, and still less well understood). We have also, at times somewhat arbitrarily, tried to make distinctions between structural/mechanistic aspects of genes and transcription and the regulation of this crucial first stage of gene expression. In presenting our case, we have deliberately omitted references from the body of the text but aimed instead to list a comprehensive set of recent reviews at the end of each chapter. From these reviews the reader can readily access the primary sources. We avoided in-text citations for two reasons: firstly, in so short a work, we felt they would unnecessarily clutter the flow of text for the reader; and secondly, it would be difficult to do justice to such a prolific research area without incorporating an inordinately large number of such citations. Transcription factor TFIIIA (see Chapter 4) is a good specific example of this problem; over the past 3 years papers on just this one particular regulatory protein have been published at the rate of about one a month, on average!

Inevitably the judgement of what is or is not an important research area is a subjective exercise; we hope we will satisfy most of our readers for most of the time. We have done our best to represent accurately what has been published at the time of writing, and any errors of fact are our responsibility and ours alone. For these, unlike tomorrow's revelations, we have no excuse!

Trevor Beebee
Julian Burke

Contents

Abbreviations	ix

1. Gene structure and transcription apparatus in prokaryotes

Gene Structure	1
General features	1
Bacterial promoters	3
Bacterial terminators	3
Other sequence elements in prokaryotic genomes	4
Transcriptional Machinery in Prokaryotes	5
Structure of RNA polymerase	5
Functions of the subunits of RNA polymerase	6
The Process of Transcription	7
DNA-binding and promoter selection	7
Closed and open polymerase complexes	7
Initiation of RNA synthesis	8
Elongation	10
Termination and anti-termination	10
Further Reading	15

2. Regulation of transcription in prokaryotes

Overview	17
Positive and Negative Control of Operons	17
Operons of prokaryotes	17
Repression of transcription (negative control)	18
Activation of transcription (positive control)	18
Mechanisms of protein–DNA interactions	19
Attenuation of transcription	21
Stringency and the Control of Growth	23
A general view of growth control	23
The molecular basis of stringency	24
Sporulation in *Bacillus subtilis*	28
Developmental changes during sporulation	28
Transcriptional regulation of sporulation	28
The role of sigma in transcription regulation	29

The Consequences of Bacteriophage Infection	30
General aspects of bacteriophage attack	30
Anti-termination after λ bacteriophage infection	30
Modification of the host RNA polymerase	31
Replacement of the host RNA polymerase	31
Further Reading	33

3. Gene structure and transcription machinery in eukaryotes

Structure and Arrangement of Eukaryotic Genes	35
General features	35
Class I genes	36
Class II genes	36
Class III genes	39
Chromatin structure	39
Transcriptional Machinery in Eukaryotes	42
RNA polymerase multiplicity	42
Structure and function of nuclear RNA polymerases in eukaryotes	42
Other eukaryotic RNA polymerases	43
Properties of nuclear eukaryotic RNA polymerases	45
Accurate *In Vitro* Transcription Systems from Eukaryotes	45
General features	45
Transcription by RNA polymerase I	46
Transcription by RNA polymerase II	46
Transcription by RNA polymerase III	47
Processing of RNA in Eukaryotes	47
Further Reading	51

4. Regulation of transcription in eukaryotes

Introduction	53
Regulation of Class I rRNA Genes in Eukaryotes	53
General aspects	53
Encystment in *Acanthamoeba*	55
Regulation of Class II Genes	55
Transcription factors in lower eukaryotes	55
Transcription factors in higher eukaryotes	56
Alternative promoters, splicing and polyadenylation	60
Regulation of Class III Genes	61
Oogenesis and embryogenesis in *Xenopus*	61
How does TFIIIA work?	64
A Rapidly Changing Scene	67
Further Reading	67

Glossary	69
Index	75

Abbreviations

aa	amino acid
ADP	adenosine diphosphate
Ala	alanine
AMP	adenosine monophosphate
Arg	arginine
Asn	asparagine
Asp	aspartate
ATP	adenosine triphosphate
β-gal	β-galactosidase
bp	base pairs
C	cytosine
cAMP	adenosine 3',5' cyclic monophosphate
CAP	catabolite activation protein
cDNA	complementary DNA
cGMP	guanine 3',5' cyclic monophosphate
d	2'-deoxyribose
DMD	Duchenne muscular dystrophy
DNase	deoxyribonuclease
Eσ	RNA polymerase holoenzyme with σ factor
f-met	formyl-methionine
G	guanine
Gln	glutamine
Gly	glycine
His	histidine
hnRNA	heterogeneous nuclear RNA
HSE	heat-shock consensus element
HSTF	heat-shock transcription factor
HSV	Herpes simplex virus I
IF2	initiation factor 2
Ile	isoleucine
IPTG	isopropyl β-D-thiogalactopyranoside
kb	kilobases (1000 bp)
Leu	leucine
Lys	lysine
Met	methionine

mRNA	messenger RNA
NTP	(unspecified) nucleoside triphosphate
NTS	non-transcribed spacer
P	purine nucleotide G or A
Phe	phenylalanine
pol	RNA polymerase
ppGpp	guanosine 5′diphosphate, 3′diphosphate
Pro	proline
Py	pyrimidine nucleotide T or C
RNase	ribonuclease
rRNA	ribosomal RNA
Ser	serine
snRNA	small nuclear RNA
SSP	starvation stringency protein
SV40	Simian virus 40
T	thymine
$t^{1/2}$	half-life
Thr	threonine
TPA	*O*-tetradecanoylphorbol 13-acetate
tRNA	transfer RNA
Trp	tryptophan
U	uracil
UAS	upstream activation sequence
X-gal	5-bromo-4-chloro-3-idolyl-β-D-galactopyranoside

1

Gene structure and transcription apparatus in prokaryotes

1. Gene structure

1.1 General features

A typical prokaryote such as *Escherichia coli* has approximately 4000 different genes, arranged along a single, circular DNA molecule and regulated in diverse ways so that the organism can respond rapidly to its environment. Unlike eukaryotes, almost all regulation is at the level of transcription; in most cases translation of the mRNA commences before transcription is completed. With the exception of genes encoding rRNA and tRNA, genes generally exist as single copies and these structural (protein-coding) sequences occupy a large proportion of the available DNA. Furthermore, and unlike eukaryotes, genes encoding enzymes forming part of a common biochemical pathway are often clustered together and co-ordinately regulated in operons. The *lac* operon is an extreme example of this, in that the three genes *Z, Y* and *A* that make up the operon are translated independently from a single (polycistronic) mRNA, each protein having its own initiation codon (AUG) and termination signal. The clustering of genes in this ordered manner (see *Figure 1.1*) therefore requires only a single regulatory switch for coordinate expression. Prokaryotic genes are co-linear, that is to say coding sequences in the DNA correspond exactly with the final amino acid sequence in the protein, and there is no interruption by intervening sequences or 'introns' (see Chapter 3).

Not all bacterial genes are expressed at all times. The activation of a gene in *E.coli* requires the binding of an RNA polymerase to a specific DNA sequence ('promoter'), usually immediately upstream of the gene(s) it controls. This interaction defines where RNA synthesis starts and also the frequency with which the gene is transcribed, ultimately determining the amount of gene product produced.

/ Gene structure and transcription

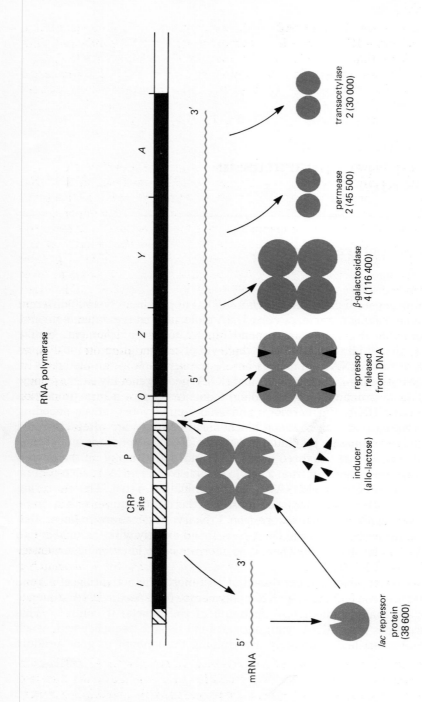

Figure 1.1. The *lac* operon of *E.coli*. The repressor encoded by *lacI* forms a tetramer which binds to *lacO* and prevents transcription. Induction is caused by an inducer, such as allo-lactose or isopropyl β-D-thiogalactopyranoside (IPTG), which binds to the repressor and causes it to dissociate from the DNA. RNA polymerase can then transcribe the structural genes encoding for β-galactosidase (*lacZ*), the permease (*lacY*), and transacetylase (*lacA*).

1.2 Bacterial promoters

E.coli promoters are characterized by the conservation of DNA sequences at positions -35 and -10 ('Pribnow box') before the start of RNA synthesis (minus numbers indicate that the nucleotide is upstream of the initiation site and not transcribed; the first nucleotide copied into RNA defines $+1$). By comparing the sequences of over 100 genes it has been found that the consensus sequences are:

      at position -35 $T_{85}\ T_{83}\ G_{81}\ A_{61}\ C_{69}\ A_{52}$
      at position -10 $T_{89}\ A_{81}\ T_{50}\ A_{65}\ A_{65}\ T_{100}$

The subscripts show the percentage frequency with which each nucleotide is found at that specific position in a promoter. Sequence analysis of the DNA between these regions shows no conservation, suggesting that the distance and not the sequence composition between positions -35 and -10 is important for promoter activity. Experimental manipulations which decrease or increase the distance between the two consensus sequences reduce the activity of the promoter; the optimum distance is 17 nucleotides. The selection of genetic mutants which up-regulate or down-regulate the promoter activity has confirmed the importance of these regions; almost 90% of such mutants fall within either of the two promoter 'boxes'. Not surprisingly, very few promoters have the optimal combination of sequence and spacing. If this were the case then all would be equally strong, thus reducing the scope for positive regulation over such a system. Recent studies have shown, intriguingly, that the promoter region at -35 can be dispensed with altogether if certain nucleotides are present immediately upstream of the promoter at -10. These experiments are beginning to show the functions of these conserved sequences, which it seems may facilitate bending of the DNA during polymerase binding.

Although the pattern of promoter boxes at positions -35 and -10 is common to most prokaryotic genes, it is not universal. Other promoter sequences can also effect promoter utilization. Striking examples of divergence from this consensus are the *nif* genes of *Klebsiella pneumoniae*. These 17 or more genes encode the enzymes required for nitrogen fixation and are induced under anaerobic conditions. They lack the canonical $-35/-10$ promoter but have instead consensus blocks centred around -24 and -12; the region at -24 includes an invariant GG dinucleotide, and that at -12 an invariant GC dinucleotide. Unusually for prokaryotes, a further upstream sequence (consensus TGTNNNNNNNNNNACA, where N = any nucleotide) is also found in a number of *nif* genes such as *nifH, nifU* and *nifB*. Presumably the great divergence from the canonical $-35/-10$ is a reflection of the extreme specialization of the nitrogen fixation process.

1.3 Bacterial terminators

The termination of transcription in *E.coli* is usually signalled by a GC-rich hairpin and loop structure formed in the transcribed RNA. Terminators fall into two major classes: those dictated by sequence alone, and those modulated by a protein

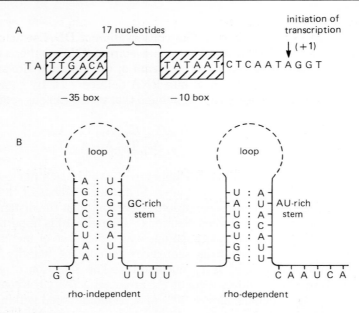

Figure 1.2. Promoter and terminator sequences in prokaryotes. (**A**) Consensus sequences for *E.coli* promoters, (**B**) rho-independent and rho-dependent terminator sequences.

factor (rho protein) largely independent of sequence. The former types have a quite well-defined consensus near the 3′-end of the gene that includes a GC-rich region of dyad symmetry sufficient to generate in the transcript a 7–10 nucleotide stem-loop some 16–20 nucleotides from its ultimate 3′-end. There is also a run of four to eight A residues in the coding strand immediately upstream of the termination point. Rho-dependent terminators lack the run of A residues, but usually have a region able to form a stable stem-loop structure.

The hairpin-loops at the 3′-end of *E.coli* transcripts may have an additional function as stabilizers of mRNA. Between 500 and 1000 copies of this repetitive extragenic palindrome are present in the chromosome and appear in about 25% of transcripts; when present they increase significantly the resistance of the RNA to degradation by 3′–5′ exonucleases.

1.4 Other sequence elements in prokaryotic genomes

Although the structures discussed above constitute the great majority (>95%) of the bacterial genome, there are certainly other important features, some of which are still being discovered. There are, for example, distinct binding sequences for repressors and for activator proteins; sites similar to terminators that can cause pausing of polymerase or effect regulation by attenuation (see Chapter 2); and other, more subtle features. An example of the latter is the 'nut-box', a 17 bp conserved sequence (including a 5 bp inverted repeat) capable of rendering RNA polymerase, transcribing through it, insensitive to rho-dependent termination; it is therefore implicated in the important regulatory phenomenon of 'anti-termination' (see Section 3.5). Some non-coding sequences of importance in bacterial DNA are depicted in *Figure 1.2*.

2. Transcriptional machinery in prokaryotes

2.1 Structure of RNA polymerase

All types of RNA in prokaryotic cells (with the exception of short primers generated during DNA replication) are transcribed by a single enzyme, namely DNA-dependent RNA polymerase. This protein has the property of polymerizing nucleotides (required in triphosphate form) into RNA, with the release of pyrophosphate, in the presence of a DNA template. The RNA product is always synthesized in the 5' to 3' direction (so the 5'-end retains a triphosphate) and is complementary to one of the two DNA strands, namely the one that runs in the opposite polarity to the transcript (the 3'−5' 'coding' strand).

RNA polymerase has been purified from many different bacteria (e.g. *E.coli, Azotobacter, Bacillus, Pseudomonas* and *Caulobacter* species) and also from blue-green algae such as *Anacystis nidulans*. The enzymes from all of these organisms demonstrate a common and highly conserved pattern of subunits; the so-called 'core' polymerase consists of four polypeptides, two α chains, one β and one β' chain. An additional subunit, σ, is usually found associated with functional enzyme to make up the RNA polymerase holoenzyme. The exact sizes of these subunits vary somewhat between species (this is especially true of the σ subunit) but in the case of the best-studied example, *E.coli*, molecular weights are now known with great precision. This is because the α subunit (mol. wt = 36 512) has been sequenced (it contains 329 amino acids) and so have the genes coding for the σ, β and β' subunits in this organism; their sizes are 70 236, 150 619 and 155 162, respectively. RNA polymerase also contains two zinc atoms per molecule, one associated with the β and the other with the β' polypeptides, all adding up to a total molecular weight of just under 450 000.

Although this pattern of subunits seems universal among the prokaryota, it was interesting to discover that it is not shared by the archaebacteria. Polymerases from several such organisms have now been purified, and although they are similar to eubacteria in that they have a single enzyme, this is usually more complex (often with several small subunits) and thus superficially more like eukaryotic enzymes. Little is known about either the mechanism or the regulation of transcription in these organisms at the present time.

It has not proved possible to crystallize RNA polymerase as yet, and attempts to investigate the three-dimensional structure of the enzyme have had to rely on methods which, at the present time, can provide only a low-resolution picture of the molecule, such as chemical cross-linking of subunits, and neutron-scattering (see *Figure 1.3*). It is generally agreed that the β and β' subunits form the central mass of the polymerase, with both of the α subunits and the σ subunit lying along the plane separating the two larger components. Conformational changes occur

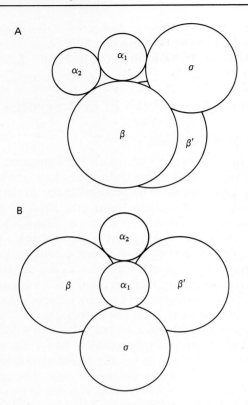

Figure 1.3. Spatial arrangements of *E.coli* RNA polymerase subunits. Views from two orientations 90° apart. (**A**) Side view, (**B**) top view.

during transcription however, associated with a shortening of the section of DNA covered by the polymerase from about 75 to 30 bp.

It is possible to reconstitute active RNA polymerase *in vitro* after dissociation of the subunits by urea, and it is also possible to look in *E.coli* cells for subunit complexes smaller than the functionally active enzyme. Studies of these complexes have supported the following scheme for the likely order of the assembly of *E.coli* RNA polymerase:

$$2\alpha \longrightarrow \alpha_2 \longrightarrow \alpha_2\beta \longrightarrow \alpha_2\beta\beta' \text{ (inactive)} \longrightarrow \alpha_2\beta\beta' \text{ (active)}$$

2.2 Functions of the subunits of RNA polymerase

The availability of mutants in polymerase subunit genes (particularly temperature-sensitive ones) as well as the stability of isolated subunits and their abilities to re-assemble into active enzyme have greatly facilitated studies on their individual roles in the transcription reaction. Even so, our understanding of subunit function is much less than complete. The role(s) of the α subunits in particular remains paradoxical. They are certainly needed for enzyme activity, but still await a specific role to be ascribed to them. The β subunit contains the active site at which nucleotides are polymerized into RNA, and is also the polypeptide to which important inhibitors of transcription, rifampicin and streptolydigin, bind. The β' subunit has a primary role in the attachment of the enzyme to the DNA template; not surprisingly, perhaps, the β' subunit is the most basic (i.e. positively charged) RNA polymerase subunit. The σ subunit is also involved in DNA-binding, this time more specifically since it is crucially concerned with the recognition of promoter sequences. It is only associated with polymerase for part of the time and can be replaced completely either by proteins with different functions, or with different σ subunits coded for by distinct genetic loci (see Chapter 2).

Recent studies on polymerase subunits are increasingly orientated towards the genes which code for them, which (not surprisingly) are co-ordinately regulated. The genes for the σ (rpoD) and α (rpoA) genes are separated from each other and from the β and β' genes on the E.coli chromosome; the β (rpoB) and β' (rpoC) genes are under the control of a single promoter and transcribed as a polycistronic mRNA. Molecular genetic and protein engineering techniques may soon enhance our understanding of RNA polymerase subunits and how they function.

3. The process of transcription

3.1 DNA-binding and promoter selection

Finding the right place to start transcription is obviously of crucial significance to the fidelity of gene expression. There are thought to be at least 2000 such promoters in the E.coli genome, buried in other DNA sequences, which amount to the equivalent of 4 000 000 non-specific binding sites. So how does RNA polymerase find the right place? Presumably simple diffusion processes are responsible for the initial contacts between enzyme and DNA, but calculations based just on diffusion cannot account for the speed with which the polymerase molecules find and associate specifically with promoter regions. It turns out that binding to non-specific DNA sequences actually speeds up promoter location by limiting the dimensions in which polymerase subsequently moves; the polymerase can then either slide along the DNA or even transfer across to another region of helix where this comes close (as it may often do in the coiled bacterial nucleoid). It has been calculated that polymerase can search some 1000 bp per second in its hunt for true promoters.

How does polymerase recognize a promoter when it finds one? Almost certainly this involves the identification, in the DNA major groove, of particular clusters of hydrogen bonds between base pairs in the so-called consensus regions of promoters, especially the 'boxes' at positions -35 and -10.

3.2 Closed and open polymerase complexes

There is now extensive evidence that the initial binding of polymerase to a promoter region is followed by substantial conformational changes in both the protein and the DNA. Taken together these changes reflect a transition from a 'closed' to an 'open' promoter complex. The initial (closed) complex yields a quite different DNA footprint from the competent (open) one, that is to say the enzyme protects different regions of the promoter from DNase digestion under the two sets of circumstances. For example, in the strong A3 promoter of bacteriophage T7 *E.coli* RNA polymerase protects positions from about -56 to -5 in the initial, closed complex, with sites of enhanced DNase I cleavage at 10 bp intervals between positions -36 and -56. On transition to the open complex, protection extends downstream to about position $+20$ and the enhanced DNase sensitivity pattern also changes. Moreover, in the closed complex the DNA remains double-stranded throughout as shown by the lack of availability of bases to various chemical modifications (especially alkylating agents which methylate bases in single-stranded DNA). Conversely, in the open complex a distinctive section of DNA (generally from positions -9 to $+3$) is 'melted out' and becomes effectively single-stranded (*Figure 1.4*). These events are underscored by studies on the polymerase–DNA complexes using cross-linking reagents, which show that in the initial (closed) complex there are usually contacts between β, β' and σ subunits and both the coding and non-coding DNA strands. Transition to the open complex results in the loss of many of these contact points and the establishment of new ones; on the *lac* promoter, for example, the σ subunit has a contact at position -3 and the β subunit at $+3$, both on the non-coding strand in the open complex. At this time the polymerase extends overall from about position -55 to $+20$, and gives maximal protection (in footprinting experiments) to the -35 and -10 box regions. In summary, events at the promoter can be summarized as:

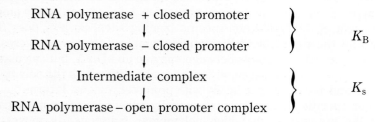

In this equation, K_B is essentially a binding constant and is generally in the range 10^7–10^9 M^{-1}; K_s on the other hand is a rate constant for the overall isomerization to the RNA polymerase open promoter complex, and normally is between 10^{-1} and 10^{-3} s^{-1}. Once formed, open complexes are very stable and decay *in vitro* with half-lives of hours rather than the minutes characteristic of closed complexes. Open complex formation is usually quicker on negatively supercoiled rather than relaxed DNA templates.

3.3 Initiation of RNA synthesis

Following the establishment of a stable open complex, the RNA polymerase can begin transcription of the DNA coding strand. The first nucleotide bound to the

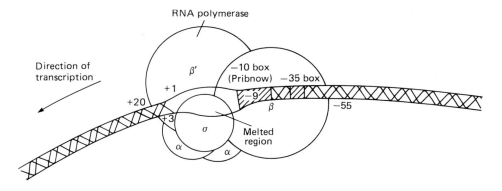

Figure 1.4. Formation of an open complex on the binding of RNA polymerase to the promoter. Local melting and bending of the DNA around the promoter region during open complex formation by RNA polymerase.

enzyme–DNA complex is almost always a purine; comparison of 88 promoters showed that 51% of nascent RNA molecules started with A, 42% with G, 5% with C and 2% with U. There are probably two distinct binding sites for nucleotides, both on the β subunit, and they are somewhat analogous with the A and P sites for polypeptide synthesis on the ribosome. An initiation site with a K_m for purines some 10-fold higher than the K_m for nucleotides (all four) at the second, so-called polymerization site has been identified.

Critical to the modern definition of initiation was the discovery of abortive initiation, an ability of RNA polymerase to generate short, mainly 2-, but up to 9-residue oligonucleotides continuously, release them and start again without ever moving from the proximity of the promoter. Abortive initiation continues indefinitely if the polymerase is supplied with less than all four nucleotide substrates, providing these include the two coded for by the first 2 bp of the gene. Indeed, it occurs as a minor reaction even in circumstances favourable to full-length RNA synthesis (i.e. in the presence of high concentrations of all four nucleoside triphosphates). RNA polymerase will also use dinucleotides of the general form pppXpY to initiate transcription *in vitro*, again providing the bases correspond to the first two coding requirements of the gene under study. This is now recognized more generally as an 'initiation mode', in which the polymerase is not yet committed to full RNA synthesis. The σ subunit remains associated with the transcription complex during the initiation mode, the cessation of which probably varies between different genes but usually comes only after some 9–16 phosphodiester bonds have been formed. Initiation is also the period during which transcription is susceptible to irreversible inhibition by rifampicin antibiotics, compounds which bind near the initiation site but are precluded after the first one or two phosphodiester bonds are formed.

The efficiency of escape from initiation mode into full elongation is probably of major importance in determining the level at which a gene is transcribed.

Typically, the half-time of open promoter complex formation is about 12 s whereas that of commencement of elongation is about 1 min; escape from the abortive initiation loop may therefore be the major rate-limiting event in RNA synthesis. On very strong promoters, however, the minimal promoter 'clearance time' may be as low as 1–2 s permitting a maximum initiation frequency close to one RNA chain per second.

3.4 Elongation

Escape from initiation into the elongation mode is accompanied by some major changes in the structure of the transcription complex. These changes are reflected in several ways; in biochemical terms the complex now becomes very stable, resistant to dissociation by concentrated salt solutions, heparin, cleavage by trypsin and relatively high temperatures. More profoundly, the σ subunit is released when the nascent RNA is 8–9 nucleotides long leaving the core enzyme to carry out all the remaining polymerization steps. The DNA remains melted out over a length of about 17 bp, in a transcription 'bubble' which moves along the helix with the polymerase. Cross-linking studies have shown that short (early) transcripts are in close proximity to the σ subunit, but as the complex processes into elongation mode the RNA becomes mainly associated with the β subunit and the DNA, and, later still, mainly with the β and β' subunits. The nascent RNA base-pairs transiently with the coding DNA strand over a stretch of about 12 nucleotides immediately upstream of the point of polymerization activity. *Figure 1.5* depicts such a transcription complex. Polymerization is achieved by nucleophilic displacement of the phosphate by ribose 3'-hydroxyl groups.

Elongation rates vary considerably depending on the nature of the template being transcribed. Maximum values range from 30 to 60 nucleotides polymerized per second, but average rates can be lower. In the case of T7 bacteriophage,

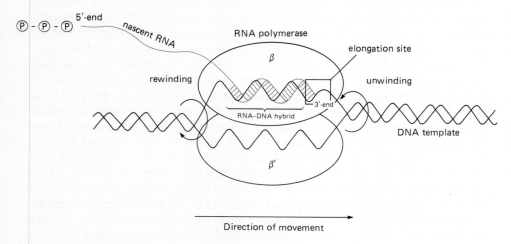

Figure 1.5. *E.coli* transcription elongation complex. DNA unwound in a transcription 'bubble' during elongation of a nascent RNA chain.

for example, the early genes are transcribed overall at 17 nucleotides per second. On this basis a typical protein-coding sequence would be copied into RNA in about 1 min. Supercoiling of DNA has no effect on elongation rates. Non-uniformity of elongation rates does however become extreme at so-called 'pause sites', regions of DNA where movement drops to 0.1 nucleotides per second. These sites are GC-rich and have structural similarities to true terminators.

3.5 Termination and anti-termination

As discussed earlier, two major kinds of termination mechanisms have been identified in *E.coli*: those which just use a signal sequence (such as the ribosomal genes), and those which require an additional protein factor known as rho (ϱ) factor. In general, what seems to happen in the former situation is that the generation of an RNA stem-loop structure causes the polymerase to slow down in a pause (though just why it should have this effect is not clear), and the subsequent incorporation of a run of U residues base-pairs so weakly with the run of A residues in the template that the transcript is released, followed more slowly by the polymerase (with half-times of dissociation of 3 and 12 min, respectively, on T7 early genes).

The ϱ factor is a protein of molecular weight 46 094 that exists *in vivo* as a hexamer and is absolutely required for healthy growth of *E.coli* cells. It is an enzyme, with an ATPase activity manifest only in the presence of (and subsequent to the binding of) single-stranded RNA at least 80 nucleotides long. An idea of how the ϱ factor functions is shown in *Figure 1.6*. Why does the ϱ factor not cause premature termination of all transcripts? Probably the answer with respect to mRNA relates to their protection, very early on in most cases, by ribosomes which initiate translation while transcription is still in progress. Only after translation termination and ribosome release, when naked mRNA becomes accessible to the ϱ factor, can termination be effected. This would explain quite nicely why such genes do not need a specific signal for transcription termination, but begs the question as to how genes making RNA and not protected by ribosomes (such as the rRNA genes) escape early termination by the ϱ factor. This in turn leads to another and final area of the transcription process, anti-termination.

Anti-termination first came to light in studies with λ bacteriophage. Immediate–early genes (those expressed first following infection of *E.coli* by the bacteriophage) of this virus are transcribed in the usual way by the host bacterium's RNA polymerase, initiating at two promoters known as P_R and P_L and terminating at the ϱ-dependent sites t_{L1} and t_{R1}. However, the RNA initiated from P_L codes for the so-called N protein, an anti-termination factor which causes the polymerase to read through both terminators into the delayed–early genes beyond (see *Figure 1.7*). How does the N protein work? The protein is unstable, with a half-life of only a few minutes *in vivo*, and needs to be synthesized continuously to maintain anti-termination. Moreover, it is specific for the t_{L1} and t_{R1} terminators, but not by virtue of sequence information in the immediate vicinity of these terminators. In fact, the information

12 Gene structure and transcription

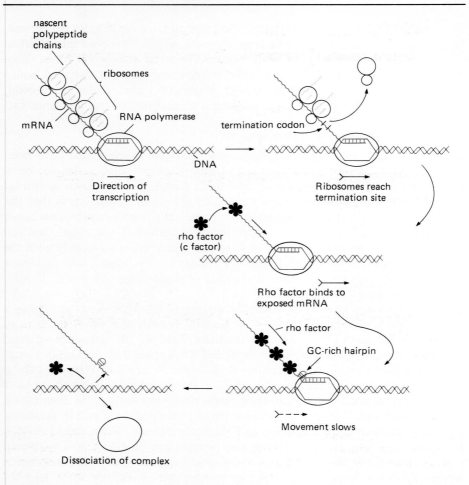

Figure 1.6. Action of ϱ termination factor. The ϱ hexamer binds to nascent RNA when space becomes available following termination of translation; it then moves to the stalled transcription complex and completes the termination reaction.

dictating susceptibility to anti-termination lies in sequences elsewhere which may be either upstream or downstream of the coding regions of the immediate – early genes; these are the nut (*N-ut*ilizing) boxes. RNA polymerase passing across a nut box is rendered liable to anti-termination when it arrives at its proximal ϱ-dependent stop site. At least one host-coded protein, *nusA*, is also involved in this process. The *nusA*, N protein and RNA polymerase probably associate as an 'anti-termination complex' yet to be clearly identified. However, *nusA* turns out to be a termination factor in its own right and can act co-operatively with ϱ at sites generally considered to be ϱ-independent (such as the t_{L2} and t_{R2}, also on bacteriophage λ DNA). Not only that, but *nusA* seems to compete

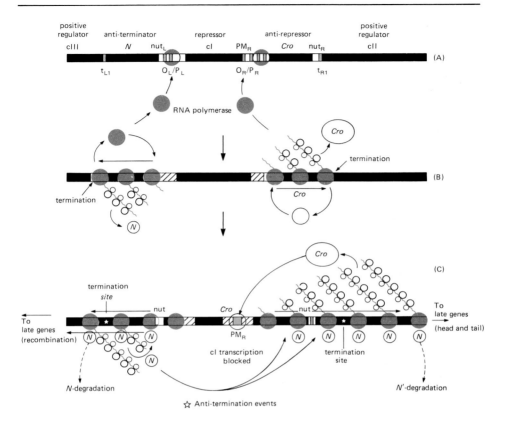

Figure 1.7. Anti-termination in λ bacteriophage lytic cycle (early phases). RNA polymerase transcribes the λ bacteriophage early genes *N* and *Cro* (**A**), (**B**), producing N protein from the former. N protein combines with RNA polymerase forming a complex capable of reading through the *N* and *Cro* termination sites, which thus proceeds to the transcription of more λ genes. The Cro protein represses transcription of the cI gene during lytic infection. P_R, *Cro* promoter; P_L, *N* promoter; PM_R, repressor promoter; O_L, *N* operator; O_R, repressor operator.

with the σ subunit for binding to RNA polymerase core enzyme (it does not bind to holoenzyme) and probably displaces the σ subunit at some stage in the transcription cycle, perhaps on most, if not all, genes. Indeed, *nusA* should probably be regarded as much a part of RNA polymerase as the σ subunit, simply associating with core at a different stage of the reaction process (see *Figure 1.8*). *nut*-A boxes are present in ribosomal genes, presumably as part of a mechanism preventing early termination by the ϱ factor.

Finally, termination is of crucial importance in the regulation of genes subject to attenuation control. The classic example of attenuation is the *trp* operon, and this is described in Chapter 2.

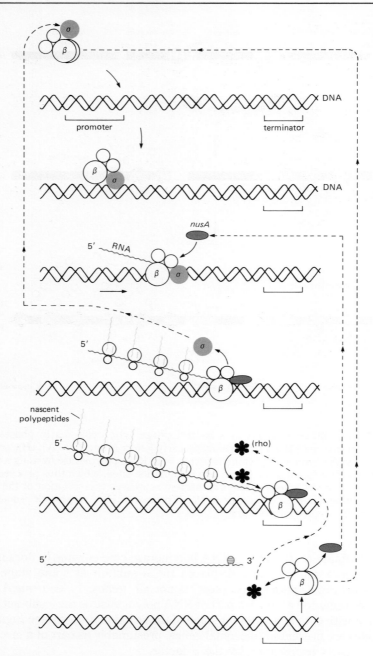

Figure 1.8. Sigma and *nusA* as alternative RNA polymerase subunits. The σ subunit is a functional part of RNA polymerase during association with DNA and transcription initiation. Early during the elongation reaction, the σ subunit is released and may be replaced by proteins such as *nusA* which affect termination specificity. The σ subunit and *nusA* proteins therefore alternate as RNA polymerase subunits.

4. Further reading

Anderson,J.E., Ptashne,M. and Harrison,S.C. (1987) Structure of the repressor – operator complex of bacteriophage 434. *Nature,* **326**, 846.

Buck,M., Miller,S., Drummond,M. and Dixon,R. (1986) Upstream activator sequences are present in the promoters of nitrogen fixation genes. *Nature,* **320**, 374.

McClure,W.R. (1985) Protein – nucleic acid interactions in transcription: a molecular analysis. *Annu. Rev. Biochem.,* **54**, 171.

Platt,T. (1986) Mechanism and control of transcription initiation in prokaryotes. *Annu. Rev. Biochem.,* **55**, 339.

Travers,A.A. (1987) Transcription termination and the regulation of gene expression. *Crit. Rev. Biochem.,* **22**, 181.

von Hippel,P.H., Bear,D.G., Morgan,W.D. and McSwiggen,J.A. (1984) Structure and function of *E.coli* promoter DNA. *Annu. Rev. Biochem.,* **53**, 389.

2

Regulation of transcription in prokaryotes

1. Overview

More than half a century's intensive research in bacterial genetics has led to an impressive, though far from complete, understanding of both the mechanics of gene expression and many aspects of its regulation. In this chapter some of the most thoroughly explored examples of gene regulation are recounted; they are thought to be representative of the ways in which bacteria regulate their genes, but of course there remains plenty of scope for, and every expectation of, the discovery of novel mechanisms in future research. It would also be wrong to leave the impression that prokaryotic genes are always regulated at the DNA level. There are now well-studied instances in which this is not the case, the best examples being the ribosomal proteins, which are autogenously regulated (i.e. their level of expression is dictated by the amounts already present in the cell) at the translation stage. Such instances are, however, very much the exception and regulation at the transcriptional level is far more common.

2. Positive and negative control of operons

2.1 Operons of prokaryotes

The structure and organization of genes in prokaryotes have been described in Section 1 of Chapter 1. In trying to understand the regulation of transcription in prokaryotes only a relatively small number of operons have been studied in detail. The *lac* operon is an excellent example of a system which has been studied extensively and found to be under negative control. A small inducer molecule is required to stimulate transcription by releasing repression. This operon also exhibits the so-called CAP system for positive regulation by cyclic AMP (cAMP).

The *lac* operon consists of three structural genes: *lacZ* (coding for β-galactosidase), *lacY* (coding for a permease) and *lacA* (coding for a transacetylase); in addition there is a regulatory region, the *lacI* gene, coding for a repressor protein, with a molecular weight of 38 000, which in its active form binds as a tetramer to the *lacO* (operator) preventing transcription of the other three genes. Neither *lacO* nor *lacP* (the promoter, containing the canonical -35 and -10 promoter boxes described in Chapter 1) encode proteins and both exert their effects in *cis* (i.e. are on the same strand of DNA).

2.2 Repression of transcription (negative control)

Mutations in *lacI* result in defective repressor and thus in constitutive expression of the structural genes even without an inducer such as allo-lactose being present. They are usually recessive, being complemented by wild-type *lacI* genes. Operator mutants are dominant, act in *cis* and also result in constitutive expression, being unable to bind repressor.

These genetic effects can now be explained in biochemical terms. Repressor is normally present at about 40 copies per cell and has an affinity for operator DNA some 4 000 000 times greater than for random sequences. Compensating for the total amount of DNA in the bacterial genome, repressors are 20 times more likely to bind at operators than anywhere else. Mutations in *lacI* which reduce repressor affinity for operator by more than 20-fold thus result in constitutive expression of the *lac* operon. Mapping such mutants also identifies the DNA-binding region of the repressor; they all reside near the N-terminal end of the protein subunits, whereas mutations near the C-terminus result in the subunits being unable to aggregate as tetramers. Mutants unable to bind inducer occur in clusters throughout the molecule; inducer-binding causes a conformational change resulting in rapid dissociation from the operator DNA.

The 26-nucleotide operator has structural symmetry in the form of an inverted repeat between positions -8 and $+28$, a generally common feature of protein-binding sequences that is also seen, for example, with activators and most restriction enzymes. Nucleotides between positions -5 and $+21$ are 'protected' (e.g. from DNase digestion) by repressor binding, and mutations acting in *cis* to give constitutive expression are found around the axis of symmetry between positions $+5$ and $+17$.

Repressor binds to the 26 nucleotide operator sequence located between -5 and $+21$ and could block transcription in two possible ways: either by inhibiting movement of polymerase bound to promoter regions upstream, or by preventing polymerase binding in the first place. Experiments *in vitro* have shown that operator saturated with repressor will not bind polymerase and, conversely, if polymerase binds first then repressor is excluded. These experiments indicate that the second mechanism is of greater importance.

2.3 Activation of transcription (positive control)

The *lac* operon cannot be induced by allo-lactose (or the synthetic analogue isopropylthiogalactoside, IPTG) in the presence of glucose, the latter compound being metabolized in preference and exerting what is known as catabolite repression of the *lac* and other specific operons. This repression can, however, be relieved by mutants in either of two other genes, adenyl cyclase (*cya*⁻) or catabolite activator protein, CAP (also known as CRP, cAMP receptor protein). The molecular basis of these observations is now well understood.

The concentrations of cAMP, produced by adenyl cyclase, in *Escherichia coli* are inversely related to glucose concentration; only when glucose is low does the concentration of cAMP increase, the cAMP then binds to CAP and positively activates the *lac* operon. The CAP–cAMP complex (CAP*) binds to sequences between -72 and -52 relative to the first nucleotide of the *lac* operon; the consensus sequence for CAP* binding is TGTGA, which often forms part of a region of dyad symmetry with the inverted sequence TCANA. How does CAP* work as a positive regulator? The binding site for CAP* varies in position from one operon to another; in the case of the *gal* operon it is between -50 and -23, and in the *ara* operon it is from -107 to -78. Its position is much more variable than the positions of the consensus promoter sequences at -35 and -10. The CAP protein could give rise to changes in DNA structure, or alternatively interact directly with RNA polymerase or other proteins. Evidence for the latter comes from studies on the λ bacteriophage repressor cI, which stimulates its own transcription by interacting directly with RNA polymerase. Some mutants of cI no longer interact with RNA polymerase and selection of second site revertants gives rise to a class of altered RNA polymerases. This mutated RNA polymerase now interacts more strongly with the CAP protein at the *lac* operon, indicating a role for direct protein–protein interaction. Furthermore, single nucleotide deletions or insertions between the CAP-binding site and the *lac* promoter strongly affect the activation.

The *nif* genes of *Klebsiella pneumoniae* (see Chapter 1) not only have an unusual promoter structure but also require a sequence around position -100 for maximal activation. This sequence, which remains functional even if moved 2 kb upstream, is a recognition site for the *nifA* protein. Although capable of activating the operons over such large distances, its effects are abolished by introducing an extra five nucleotides (i.e. half a helical turn) between it and the promoter; adding or deleting 10 nucleotides, a full turn, has little effect. It seems that the *nifA* protein must present a specific aspect to the protein complex at the promoter in order to stimulate transcription.

2.4 Mechanisms of protein–DNA interactions

Many prokaryotic DNA-binding proteins, both repressors and activators, share common structural features. Central to this homology is the motif 'α-helix–turn–α-helix', in which the helices bind in the major grooves of the DNA (*Figure 2.1*). Specificity in the case of λ bacteriophage and other repressors is provided by hydrogen-bond formation between α-helix amino acid side chains and exposed positions on the DNA base pairs; thus for λ cI repressor the structure is helix 1 (eight amino acids), turn (three amino acids) and helix 2 (seven amino acids). Alanine is often conserved in helix 1, glycine in the second position of the turn, and valine in position four of the second helix.

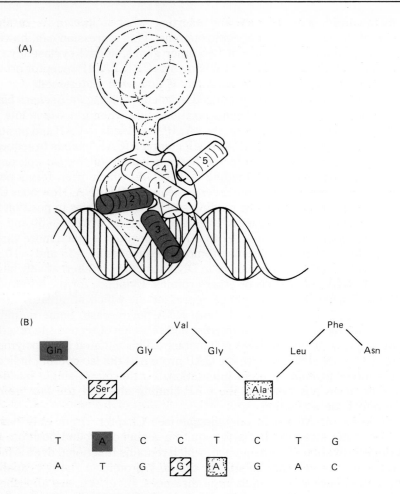

Figure 2.1. Repressor – DNA interactions in λ bacteriophage. **(A)** Representation of λ bacteriophage repressor monomer. The protein molecule is shown as a dumb-bell, with the N-terminal containing five regions of α-helix. Helices 2 and 3 are involved with DNA binding with helix 2 across the major groove and helix 3 lying in the major groove. The molecule normally exists as a dimer when bound to DNA. **(B)** Interaction of α-helix 3 of the λ repressor with DNA sequences in O_R1 of λ DNA. Glutamine (Gln) interacts with the A residue as indicated, serine (Ser) with G, and alanine (Ala) with another A residue.

For long range interactions between modulator binding sites and promoters the most likely mechanism of stimulation or repression involves direct protein–protein contacts, with intervening DNA looping out to facilitate them. For example, one section of λ bacteriophage DNA contains three adjacent binding sites for cI repressor; binding of a molecule of repressor to one site (O_R1) increases the affinity of the second (O_R2) some 10-fold. This co-operative effect

Table 2.1. Amino acid leader sequence of five biosynthetic operons regulated by attenuation

Operon	Leader sequence
his	Met Thr Arg Val Gln Phe Lys His His His His His His His Pro Asp
trp	Met Lys Ala Ile Phe Val Leu Lys Gly Trp Trp Arg Thr Ser
phe	Met Lys His Ile Pro Phe Phe Phe Ala Phe Phe Phe Thr Phe Pro
leu	Met Ser His Ile Val Arg Phe Thr Gly Leu Leu Leu Leu Asn Ala Phe Ile Val Arg Pro Val Gly Gly Ile Gln His
thr	Met Lys Arg Ile Ser Thr Thr Ile Thr Thr Thr Ile Thr Ile Thr Thr Gly Asn Gly Ala Gly

Underlining shows the likely regulatory codons.

is still seen if the sites are artificially separated by an integral number of helical turns, and electron microscopy has confirmed that in this situation the repressors are in contact with each other and the extra DNA simply loops out between them. Thus at present there is no evidence, at least for bacterial repressors and activators, to support alternative models involving untwisting DNA or the formation of Z-DNA.

2.5 Attenuation of transcription

Amino acid biosynthetic operons (see *Table 2.1*) exhibit yet another form of transcriptional regulation in which transcription terminates prematurely upstream of the structural genes of the operon, if the appropriate species of tRNA charged with the amino acid product of the enzymes encoded by the structural genes of the operon are present. An example of this type of control is the tryptophan (*trp*) operon, encoding five proteins involved in the synthesis of tryptophan from chorismate. Regulation involves a promoter, an operator and an attenuator (see *Figure 2.2*). The operator has a two-fold symmetrical axis, binds a repressor and thus prevents transcription. Unlike the *lac* repressor, however, that for the *trp* operon will only bind to the operator in the presence of its modulator (tryptophan itself).

The attenuator lies between the operator and the first structural gene sequence, and gives rise to an RNA leader that can form complex secondary structures capable of terminating RNA polymerase before it ever reaches the genes. This negative control was discovered by the observation that deletions in the leader region actually increased *trp* mRNA synthesis in the presence of tryptophan. What happens is that after initiation of transcription, ribosomes bind to and begin translation of the leader sequence, which includes a standard initiation signal for translation and two tryptophan codons. In the absence of charged Trp–tRNA (i.e. when tryptophan levels are low) the ribosomes soon dissociate, allowing polymerase to continue and transcribe the 7 kb polycistronic mRNA. If tryptophan is present, however, the ribosome reads on through and in so doing alters the base-pairing of the leader to create a stem-loop termination sequence

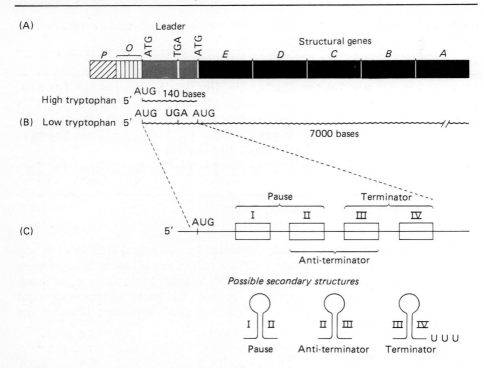

Figure 2.2. Control of the *trp* operon by attenuation. (**A**) The gene organization of the tryptophan operon showing the promoter (*P*), operator (*O*), the leader region encoding the attenuating RNA, and the structural genes *E–A*. In the presence of high levels of tryptophan a small, 140 nucleotide RNA is synthesized; in the absence of tryptophan, a large polycistronic message is produced. (**B**) Transcription is regulated by RNA secondary structure in the leader. The regions that pair to form the different regulatory complexes are indicated in (**C**).

typical of the ϱ-independent termination site (see Chapter 1). RNA polymerase then stalls and dissociates from the template together with the short (140 nucleotide) leader transcript. The RNA transcribed from the leader sequence has the ability to form three loops which act as pause, anti-terminator and terminator. In the initial stages transcription of the message starts, but very soon a secondary structure in the newly transcribed RNA forms between regions I and II (see *Figure 2.2*). It is thought that the RNA polymerase then pauses. Meanwhile a ribosome binds to the leader and initiates translation at the AUG codon. The ribosome then pulls the mRNA with the paused RNA polymerase, this releases the secondary structure between regions I and II and RNA synthesis is restored. The pausing and ribosome-induced release of RNA polymerase probably functions to synchronize transcription and translation. The next step is dependent upon the presence or absence of charged tRNAs. In the absence of a charged tRNA at a regulatory codon required for synthesis of the leader, the ribosome stalls and an anti-terminator loop forms from regions II and III.

This loop prevents a third terminating loop (between III and IV) from forming and therefore transcription continues into the structural gene.

In the presence of a charged tRNA at a regulatory codon the anti-terminator loop between regions II and III does not form and the ribosome continues translation until it reaches the leader peptide stop codon. The RNA polymerase continues transcription and as region IV is copied into RNA this forms a terminating loop by hydrogen bonding with region III. This stem-loop, followed by a run of U residues, provides a termination signal for RNA polymerase which then dissociates from the DNA.

Regulation by attenuation is a subtle mechanism. The *trp* operon has only two tryptophan codons in the leader and only comes into play during acute starvation. In contrast the *his* operon has seven codons for histidine in the leader and consequently attenuation is therefore very sensitive to the level of charged His–tRNAs; in this operon attenuation is the major mechanism by which transcription is regulated.

The *ilvGMEDA* operon is regulated by attenuation, which is controlled by valine, isoleucine and leucine tRNAs; not surprisingly all three codons are found scattered in the leader RNA. The *thr* operon which is regulated by two tRNAs (threonine and isoleucine) has eight threonine codons and four isoleucine codons interspersed in the leader sequence. This may be important in ensuring that the ribosome stalls in the same position, irrespective of the missing amino acid.

Attenuation thus acts in the same direction as repression, presumably to increase the effectiveness of suppression. The *trp* repressor by itself is much less efficient in this respect than is the *lac* repressor.

3. Stringency and the control of growth

3.1 A general view of growth control

Whereas operons represent classical examples of precise regulation of specific subsets of cellular genes, they are not by any means the only kind of mechanism of transcriptional control associated with the bacterial genome. In particular, it has been found that ribosome production, and the overall rate of protein synthesis, is strictly correlated with the doubling-time of the *E.coli* cell in culture whereas transcription of most structural genes is unrelated to this measure of growth rate. An experimental approach which has been particularly successful in the study of growth-related phenomena has been to make use of the so-called stringent response. Wild-type bacteria shut down protein synthesis rapidly when starved of amino acids, an event followed quickly (within a few minutes) by a dramatic down-regulation of the transcription of genes coding for 'stable' RNA (rRNA and tRNA); this is the stringent response, which once again is not reflected in the mRNA population in general. Stringency may be a useful model for growth control under more normal circumstances, though whether all aspects of the regulatory processes involved are identical remains to be seen.

Under normal growth conditions, transcription of stable RNA species represents the overwhelming majority of the total. Probably 40–50% of RNA synthesis is constituted by transcription of the seven rRNA genes alone, and because of their stabilities relative to mRNAs the rRNA and tRNA constitute some 97% of the total RNA in a bacterial cell. Furthermore, fusion of the promoters of the rRNA or tRNA genes and their associated DNA regions to other genes (such as β-galactosidase) renders them subject to growth-rate control. Ribosomal protein promoters do not have this effect, and, as mentioned earlier (Section 1), the expression of these genes under different growth conditions is apparently modulated at the translational level.

3.2 The molecular basis of stringency

Amino acid starvation causes stable RNA synthesis to fall some 10- to 20-fold *in vivo*, whereas bulk mRNA transcription falls no more than 2- to 3-fold. This apparent change in relative promoter utilization is accompanied by sharp increases in the concentration (from $<50~\mu M$ to >1 mM) of an unusual nucleotide, guanosine 5' diphosphate, 3' diphosphate (ppGpp). This molecule is synthesized by a ribosome-associated enzyme, the product of the *relA* gene, when the appropriate (i.e. correctly base-paired) uncharged tRNA is bound at the A site of the ribosome. GTP and possibly also GDP are phosphorylated by a molecule of ATP in a so-called 'idling reaction' possible only in the absence of the amino acid substrate for normal protein synthesis.

$$\text{GTP (or GDP)} + \text{ATP} \longrightarrow \text{ppGpp} + \text{AMP}$$

Mutations in *relA* can lead to a 'relaxed' phenotype, in which ppGpp is not synthesized. Such cells do not demonstrate stringent control of stable RNA synthesis, an observation which strongly implicates ppGpp in the regulatory mechanism. Also of interest has been the identification of relaxed mutants where the lesion is not in ppGpp synthesis, but in RNA polymerase subunits. This in turn suggests that ppGpp might exert its effects by some direct interaction with the transcriptional apparatus.

Another consequence of amino acid starvation is a profound change in the overall pattern of protein synthesis, leading to a situation where a single species, the starvation stringency protein (SSP), constitutes up to 50% of the total cell protein being synthesized. Intriguingly, the SSP forms an equimolar complex with RNA polymerase holoenzyme (it does not bind to core) and inhibits its activity *in vitro* on at least some promoters. The gene for this factor has recently been cloned and there are no promoter sequences approximating to the usual -10 and -35 consensus regions (see Chapter 1). The ppGpp and SSP aspects of the stringent response have not yet been reconciled into a general picture of events, though it will be interesting to establish the circumstances under which the SSP promoter is active *in vitro* (with ppGpp?) and of course the significance of the SSP–polymerase complex.

So how does stringency exert its selective effects on gene activity? Further insight into this problem has arisen from studying the DNA sequences around stable RNA gene promoters. Ribosomal genes are unusual in having two tandem

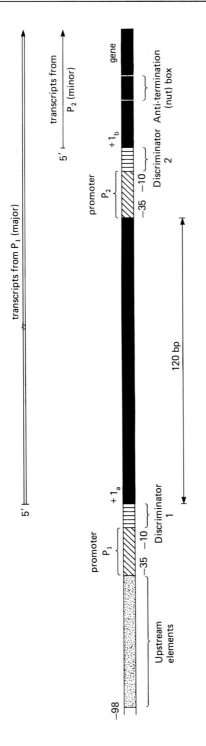

Figure 2.3. Ribosomal gene promoter. The double promoter (P_1 and P_2) arrangement upstream of *E.coli* ribosomal genes, showing discriminator elements and anti-termination box.

promoters, P_1 and P_2, separated by about 120 bp (see *Figure 2.3*). Both have classical -10 and -35 promoter boxes but they differ in other respects, notably in that P_1 (the furthest upstream) is most active during rapid cell growth, and is the only one subject to stringent control. At slow growth rates, therefore, the relative contribution of P_2 is greatly enhanced. Both P_1 and P_2, as well as tRNA promoters, are characterized by highly GC-rich sequences (consensus: 5′ GCGCCNC 3′) between positions -7 and -1; these 'discriminator' sequences are not found in most mRNA promoters, and their function remains unclear in light of the facts that P_2 has one but is not subject to stringent control, and operons for ribosomal protein have them but are apparently mainly regulated at the translational level. Even so, fusion of rRNA promoters to other genes renders them liable to stringent control and mutation in the discriminator region of the tyrosine tRNA gene abolishes stringent regulation *in vivo*; it also alters ppGpp response of the promoter *in vitro*, and it seems certain that discriminators are involved in some way in stringency. However, the most recent studies of ribosomal promoter P_1 has shown that, in this case, growth control is dependent upon sequences outside the discriminator, specifically between -21 and -51. Evidently there is some way to go before the significance of promoter sequences to growth regulation is properly understood.

The strength of stable RNA gene promoters is not related to the presence or otherwise of discriminators; regions upstream of the -35 promoter box are implicated in this property of the genes, probably involving sequences between -40 and -100 but without extensive homologies between different types of stable RNA genes. These upstream regions do however contain sequences resembling the -35 and -10 promoter boxes, and may function as loading sites for RNA polymerases.

So how does ppGpp regulate stable RNA synthesis? There is evidence from *in vitro* studies that this nucleotide can interact directly with RNA polymerase to alter its promoter selectivity, and it does so in a way which disfavours ribosomal and transfer RNA genes. The effect is reversible, specific for ppGpp (other guanosine nucleotides do not work) and seems to be dependent on the rate of open complex formation (see Chapter 1, Section 3.2), with later stages of transcription relatively unaffected. Perhaps the most likely explanation of these and other observations is that RNA polymerase is an allosteric enzyme that can (and does) exist *in vivo* in several functional forms. These isomers may be in equilibrium, differ from one another with respect to their abilities to initiate transcription from different types of promoters, and interconvertible with agents such as ppGpp. Rates of transcription from a particular gene would therefore depend upon its inherent promoter structure (including sequence elements preferred by one or more RNA polymerase isomers) and upon the relative amounts of the appropriate isomers in the cell. A continuum of responses to modulators such as ppGpp would therefore be expected, with some genes responding to lower concentrations than others, and this is what is actually found. The synthesis of tRNA, for example, is inhibited by ppGpp concentrations *in vivo* more than an order of magnitude lower than those needed to inhibit ribosomal gene activity.

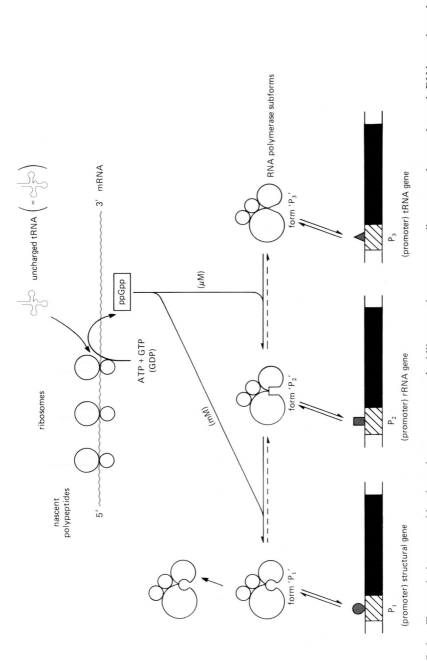

Figure 2.4. Transcription control by the stringent response. An idling reaction occurs on ribosomes when uncharged tRNAs are bound leading to the synthesis of ppGpp and pppGpp. These nucleotides may interact directly with the RNA polymerase causing conformational (allosteric) changes that favour the interaction of the RNA polymerase with promoters for structural genes (●) rather than those for stable RNA (rRNA, ■ and tRNA, ▲).

Studies *in vitro* have demonstrated effects of several other cellular macromolecules on RNA polymerase promoter specificity, including the initiator formyl-methionine–tRNA (tRNA–f-met), translation initiation factor IF2 and the elongation factor complex TuTs. The relevance of these observations to events *in vivo* remains somewhat uncertain and needs greater scrutiny by independent methods (especially the isolation and characterization of more suitable mutants). *Figure 2.4* is an overview of molecular events thought to be involved in these processes. Obviously the mechanisms underlying growth control and stringency are quite different from those regulating operons. A far broader spectrum of genes is affected during growth control, and though the details remain to be elucidated it seems likely that central to the process is allosteric modification of RNA polymerase itself.

4. Sporulation in *Bacillus subtilis*

4.1 Developmental changes during sporulation

Bacillus subtilis is a Gram-positive bacterium that can exist in two very different morphological forms; either as a typical 'vegetative' cell when growing in a rich nutrient medium, or as a 'spore'. Spores form when nutrients become depleted; they have a replicated and segregated extra genome, and a hard external coat. Sporulation takes many hours to complete and is associated with major changes (up to 40% of sporulation mRNA is spore-specific) in the pattern of gene expression.

4.2 Transcriptional regulation of sporulation

Sporulation is associated with major changes in RNA polymerase holoenzyme, the most significant of which seem to involve the σ subunit. Before considering these in detail, however, it is necessary to describe the situation in vegetative cells because it turns out to be more complex than in *E.coli* (where to all intents and purposes only one σ subunit, σ^{70}, occurs). During normal growth probably more than 90% of *B.subtilis* RNA polymerases contain σ^{43}; this leaves a small but significant minority with different σ subunits of which there are at least three: σ^{37}, σ^{32} and σ^{28} (where the numbers represent estimated molecular weights, in kilodaltons). It has been firmly established that these minor σ subunits are not partial breakdown products of σ^{43}, and indeed share little or no amino acid sequence homology with it. The holoenzyme with σ^{37} (Eσ^{37}) recognizes a characteristic subset of the -10 and -35 promoter boxes, that is a distinct set of promoters; these include at least two genes involved in the onset of sporulation and several others particularly active about that time, including that for the secreted protease subtilisin and, interestingly, one of three tandem promoters for the gene for σ^{43}. Thus, although Eσ^{37} is found in vegetative cells its activity seems to peak at the end of log phase and to be in some way involved in the onset of sporulation. The promoters recognized by Eσ^{32} are different again, and are often found paired with promoters used by other holoenzymes (especially $E\sigma^{37}$) forming promoter tandems on, for example, the sporulation gene *spoVG* and subtilisin. Finally, $E\sigma^{28}$ also recognizes its own sets of canonical -35 and -10 promoter sequences, and probably transcribes 25–30 genes in *B.subtilis*. The nature of these genes remains largely unknown, though one promoter is probably a heat-shock locus for σ^{43} and, in general, the promoters for the σ^{28} gene seem to be developmentally down-regulated at the onset of sporulation.

The progression of sporulation is associated not just with changes in activities of pre-existing holoenzymes but also with the appearance of at least one, probably two, new ones. The *spoIIG* gene is activated early in the differentiation process and codes for the σ^{29} subunit (with an actual mol. wt of 24 000, see below), which displaces the σ^{43} subunit on the bulk of the cell's holoenzyme population. The $E\sigma^{29}$ holoenzyme transcribes a wide range of sporulation-specific genes, using yet again characteristic promoters with distinctive -35 and -10 promoter boxes. A further sporulation gene product (*spoIIAC*) has also been tentatively identified as a σ subunit on the basis of amino acid sequence homology, but not yet by functional tests. Unlike the vegetative minor σ subunits, both of the sporulation σ subunits show significant sequence homologies with σ^{43}.

4.3 The role of sigma in transcription regulation

These examples amply demonstrate one more mechanism for transcriptional regulation, notably the replacement of one σ subunit by another capable of conferring distinctly different properties on the RNA polymerase as a whole. It may be too soon to generalize, but replacement of the σ subunit seems to be a mechanism used in response to environmental deterioration and subsequent stresses. Most of the genes regulated by changes in the σ subunit are non-essential during normal growth conditions. Interestingly, it has now been shown that *E.coli* also can replace its usual σ subunit, this time during heat shock. Brief exposure of cells to temperatures above that at which they normally grow induces a heat-shock response, involving the switching off of most genes but relative increases in the activities of 17 specific ones. In *E.coli* one heat-shock locus, *HtpR*, codes for a new σ subunit (σ^{32}; mol. wt = 32 000) which displaces the usual sigma subunit on the holoenzyme (σ^{70}) and allows it to recognize distinctive heat-shock gene promoter sequences. As in the case of *B.subtilis*, one of these is a regulator of transcription of the gene for the σ^{70} subunit. The *E.coli* σ^{32} subunit has a very short half-life *in vivo*, decaying only a few minutes after triggering the heat-shock response. Other examples of changes in σ subunits are also coming to light, and a particularly interesting one promises to be the involvement of new σ subunits in nitrogen fixation, notably σ^{60} (the product of the *ntrA* gene).

One final point about confusion in the literature is that generally σ proteins all seem to exhibit anomalous migration on denaturing gels, which routinely overestimate their molecular weights. The values given here are in most cases 'real' because they have been calculated from gene sequencing results; they are usually about 80% of estimates calculated from mobility on polyacrylamide gels. Older publications give the sizes of σ subunits in *E.coli* and *B.subtilis* of 85 000–90 000 and 55 000, respectively.

5. The consequences of bacteriophage infection

5.1 General aspects of bacteriophage attack

The invasion of a bacterial cell by bacteriophage viruses precipitates a series of dramatic changes in RNA and protein biosynthesis, the nature of which vary considerably depending upon the type of bacteriophage. These changes reflect the competition for survival between two quite different genomes, and cannot therefore be used as simple models of regulatory phenomena in healthy cells. Nevertheless, bacteriophage infection is a valuable experimental tool and has generated a number of insights into mechanisms which, it increasingly turns out, are also employed by bacteria under more normal circumstances.

Lytic infection is usually a rapid process, leading to destruction of the host cell and the release of replicated bacteriophage particles in less than an hour. During this period, a subset of genes (designated early or immediate – early) on the bacteriophage DNA are transcribed by an unmodified bacterial RNA polymerase. These genes have promoters homologous to those found on the bacterial chromosome (though they are usually strong promoters) and are expressed even in the absence of host protein synthesis; no new factors are required. Invariably one or more of these early genes codes for products which modify the transcription apparatus, in such a way as to select for transcription of the remainder of the bacteriophage genome and against continued transcription of the bacterial DNA. These bacteriophage genes expressed at later times of infection may themselves be further subdivided, into 'middle' and 'late', for example, primarily on the basis of the time scale over which their transcripts become detectable in the cell. A common feature is that they require protein synthesis, using the translation machinery of the host, for their own expression; this allows translation of mRNA from bacteriophage early genes and the production of protein factors which are the agents of transcriptional modification. There may be cascades of consequential effects (e.g. with σ subunits following SPO1 infection of *B.subtilis*), and in any case it is during the later phases of infection that the major changes are seen. By the time the bacteriophage DNA is replicating and the bacteriophage coat proteins are being synthesized, host gene expression has usually ceased almost completely.

5.2 Anti-termination after λ bacteriophage infection

It is in the transition from early to middle – late bacteriophage gene expression that the interesting effects on the transcription apparatus are observed. Bacteriophages differ widely in their adopted strategies for changing selectivity of RNA polymerase from host to invader. In λ bacteriophage an immediate – early gene product, the N protein, causes anti-termination allowing RNA polymerase to 'read through' into the delayed – early genes. In fact these delayed – early genes code for, among other things, another anti-termination factor (Q) which permits further read through into the late genes. Anti-termination is therefore an important aspect of infection by λ bacteriophage (see also Chapter 1).

5.3 Modification of the host RNA polymerase

Other strategies employed by bacteriophages include various modifications of the host RNA polymerase to alter its promoter selectivity. SPO1 does this by replacing the *B.subtilis* σ subunit with products of its own genome. *E.coli* bacteriophage T4 employs a fundamentally similar approach. Expression of T4 early genes (transcribed by normal *E.coli* RNA polymerase) leads to a number of subsequent modifications to the host enzyme: α subunits are ADP-ribosylated, and five small (mol. wt <30 000) bacteriophage-coded polypeptides become associated with the core polymerase. Three of these proteins (products of T4 genes 33, 45 and 55) are essential for the transcription of T4 bacteriophage late genes and one in particular (gene 55 protein, mol. wt 23 000) apparently replaces the normal σ^{70} subunit. Polymerase altered in this way selects for transcription of actively replicating DNA templates (overwhelmingly the bacteriophage DNA in late infection) and against the host DNA.

5.4 Replacement of the host RNA polymerase

Perhaps the most dramatic effects of infection are seen after invasion by one of the smallest types of bacteriophage, notably the closely related T3 and T7 bacteriophages (see *Figure 2.5*). These short, linear DNA viruses have less than 25 genes subdivided into early or class I (five genes), middle or class II (seven genes) and late or class III (12 genes). Class I genes are clustered in the left-hand 20% of the genome and are transcribed by the unmodified RNA polymerase, as normal, 4–8 min after infection. There are five proximal and strong promoters (A1, A2, A3, B and C) which give rise to polycistronic mRNAs (immediately cleaved by RNase III) which in turn are translated by *E.coli* ribosomes. Gene one of this complex codes for a totally new RNA polymerase, a single-subunit enzyme resistant to rifampicin, lacking zinc atoms, and with a molecular weight of 99 000 (883 amino acids). The bacteriophage-specified enzyme is responsible for the transcription of both class II genes (6–15 min post-infection) and class III genes (7 min post-infection through to cell lysis). The 17 promoters for bacteriophage class II and III genes are quite unlike classical *E.coli* (or early gene) promoters, but similar to each other; they include a 23 bp highly conserved sequence extending from position −17 to +6 which is very GC-rich. T7 bacteriophage polymerase is highly specific for these promoters, but binds with relatively low affinity ($K_{ass} = <10^7$ M^{-1}). About 8 bp of DNA are melted during the binding reaction, which seems generally analogous to the process observed with bacterial polymerase and its cognate promoters. The only difference between class II and class III gene promoters appears to lie in the AT-richness of upstream regions, which tends to be greater in the latter than in the former. Another impressive feature of T7 bacteriophage RNA polymerase is its rapid elongation rate, which at 200 nucleotides per second is four to five times faster than the maximum reported for the *E.coli* enzyme.

As far as we know, this total replacement of one polymerase by another has no analogy in the healthy bacterium. It is an extreme case, but one of increasing use to molecular biologists. The insertion of genes downstream from bacterio-

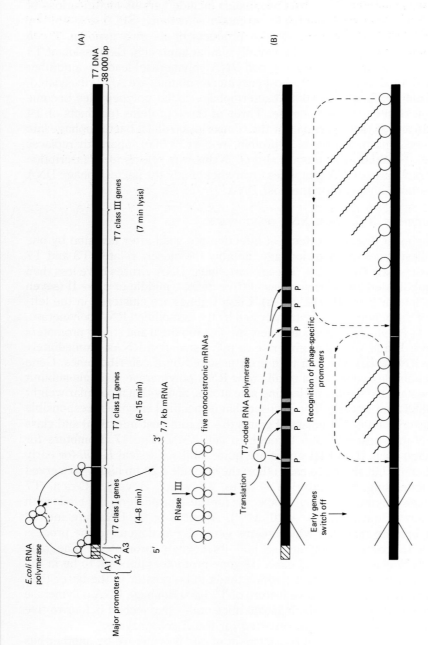

Figure 2.5. T7 bacteriophage: the molecular biology of infection. Immediately after T7 infection, early (class I) genes are transcribed by host bacterial RNA polymerase from strong promoters clustered at the left-hand end of the genome (**A**). One of the T7 class I gene products is a novel RNA polymerase that recognizes only T7 promoters (p) located in the T7 class II or III gene regions, and transcribes these late genes as the infection proceeds (**B**).

phage promoters such as T7 (or, more commonly, bacteriophage SP6) has facilitated the generation of clones from which very large quantities of any desired transcript can be synthesized *in vitro*. This is useful for nucleic acid hybridization and related studies on the structure and function of many genes from a variety of organisms.

6. Further reading

Doi,R.H. and Wang,L.F. (1986) Multiple prokaryotic sigma factors. *Microbiol. Rev.*, **50**, 227.
Lamond,A.I. (1985) The control of stable RNA synthesis in bacteria. *Trends Biochem. Sci.*, **10**, 271.
Ptashne,M. (1986) *A Genetic Switch*. Cell Press and Blackwell Scientific Publications, Oxford.
Yanofsky,C. (1987) Operon-specific control by transcription attenuation. *Trends Genet.*, **3**, 356.

3

Gene structure and transcription machinery in eukaryotes

1. Structure and arrangement of eukaryotic genes

1.1 General features

The organization of genes in eukaryotes differs from that in prokaryotes in several major respects. At the gross level, eukaryotic genes are distributed on a number of chromosomes rather than one. This has important consequences for regulation as regulated genes (e.g. the human α- and β-globins) are often located on different chromosomes. The concept of the operon does not seem to apply in eukaryotes and polycistronic mRNA is rare or non-existent. Furthermore, eukaryotic genes are often interspersed between an excess of non-coding DNA sequences. Even the simplest eukaryotes are more complex genetically than prokaryotes; *Drosophila* or sea urchin genomes probably contain up to 20 000 genes, and mammals up to 100 000, as estimated by genetic saturation studies and nucleic acid hybridization analyses. Another important difference is that, in contrast to *E.coli*, eukaryotic protein-coding genes are often split and the coding 'exons' are interspersed with (generally) non-coding 'introns'. A consequence of this is the enormous size of some eukaryotic genes, such as that responsible for the human genetic disease Duchenne muscular dystrophy (DMD). The DMD gene may cover 2000 kb of DNA, one thousand times bigger than the *E.coli lacZ* gene and equivalent to one half of the *E.coli* genome. Finally, the nuclear membrane in eukaryotes effectively separates the processes of transcription and translation in both space and time.

Unlike prokaryotes, eukaryotes have three types of RNA polymerase each of which is responsible for the transcription of different classes of genes. Class I genes are those coding for the large rRNAs (18S and 28S in mammals), and also 5.8S RNA, in the nucleolus; class II genes include all those coding for proteins, and some small nuclear RNAs (snRNAs); and class III genes include the tRNA genes, 5S rRNA genes and genes for some snRNA species.

1.2 Class I genes

Ribosomal RNAs are extremely abundant; a mammalian tissue-culture cell dividing every 24 h may contain as many as 10 000 000 ribosomes. High levels of rRNA synthesis are possible because cells have multiple copies (several hundred) of the 6–15 kb gene coding for a single precursor containing (in mammals) 18S, 5.8S and 28S rRNAs. Each active gene may carry 100 or more molecules of RNA polymerase I at any one time. These reiterated genes are clustered in tandem in the nucleolus, separated by a region mistakenly called the non-transcribed spacer (NTS). Core promoters on rRNA genes have been defined as about 45 bp long and between positions −35 and +9. Regions up to −150 bp upstream may also influence transcription rates, but to an extent very variable between different species. Promoters for the rRNA gene generally tend to be much more species-specific than those for class II genes. In *Xenopus* the NTS contains multiple copies of two sequence elements: the first, spacer promoters, are highly homologous with the major promoter while the second is a 42 bp enhancer, which is a sequence that can stimulate gene transcription in a bi-directional manner. It is now clear that transcripts initiate from these spacer promoters and all terminate at a site immediately upstream of the major promoter, a mechanism apparently designed to deliver large numbers of RNA polymerase molecules to the site of initiation of the precursor molecule. It turns out that in *Xenopus* even polymerases that have transcribed the functional gene continue through the next NTS, generating a rapidly degraded RNA, and terminate at the same site adjacent to the promoter of the next gene. These terminators (at ∼position −168) actually overlap the promoter and include an 18 bp tandem repeat (*Figure 3.1*). In mice, however, there seems to be a true termination site about 500 bp downstream of the 3′-end of the gene so the RNA polymerases may not traverse the NTS in the same way.

1.3 Class II genes

As with *E.coli*, most eukaryotic protein-coding genes contain the canonical TATA sequence in their promoters. In yeast a consensus TATAAA is found between positions −40 and −120, but unlike both prokaryotes and higher eukaryotes deletion of this region does not affect either the position of initiation or the amount of transcript produced. It has been suggested that these determinants in yeast may reside in some other sequence between positions −91 and +1. Multicellular eukaryotes have a TATA box at around −25, with the consensus:

$$T_{82} \; A_{99} \; T_{93} \; A_{83} \; \begin{matrix}T_{37}\\A_{63}\end{matrix} \; \begin{matrix}T_{33}\\A_{83}\end{matrix} \; A_{50}$$

Deletion analysis with the β-globin gene shows that removing this sequence reduces transcription 20-fold *in vitro*; changing TATA to TAGA or TAAA in the conalbumin gene also abolishes transcription. The TATA box determines the precise initiation site, as shown by cloning the region −32 to −12 from the adenovirus major late promoter (a very strong promoter) into a bacterial plasmid;

Gene structure and transcription machinery in eukaryotes 37

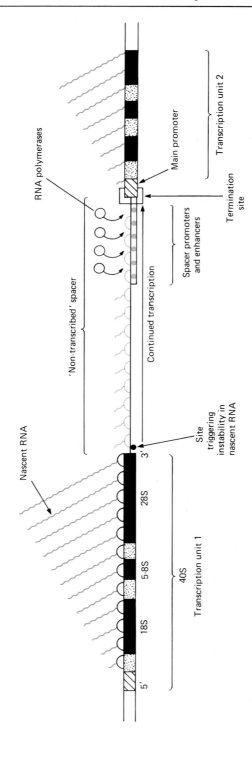

Figure 3.1. Class I (nucleolar ribosomal) gene structure in *Xenopus*. Tandemly linked ribosomal genes separated by spacer DNA containing spacer promoters, enhancers and a termination site immediately upstream of the next main promoter.

transcription then initiates 30 nucleotides downstream *in vitro*, on bacterial DNA, in the presence of eukaryotic transcriptional components.

Increasing numbers of higher eukaryotic promoters are being found which lack a TATA box, but instead have a GC-rich region (such as GGGGCGGAGC in the hypoxanthine phosphoribosyltransferase gene) centred around position -33. These may be a feature of 'housekeeping' genes, that are expressed in all cells; this situation also occurs in dihydrofolate reductase and phosphoglycerate kinase, for example, and another interesting feature of them is that they often generate transcripts with multiple 5'-ends. The snRNAs transcribed by polymerase II also lack TATA boxes.

Deletion analyses of several yeast promoters have shown that sequences hundreds of nucleotides upstream are required for transcription. These upstream activation sequences (UAS) such as UAS1 and UAS2 at positions -265 and -229 in the cytochrome *c* oxidase (*CYC1*) gene bind specific proteins that influence transcription. In contrast, some yeast sequences behave in *cis* as 'silencers'; thus in the yeast mating locus a DNA sequence termed HMRE can reduce the activity of promoters even when 2.5 kb away. In higher eukaryotes, the first recognized upstream sequence was CAAT at about -70. This is crucial for the *in vivo* expression of transfected genes, but non-essential for transcription *in vitro*. It is not usually found in housekeeping genes lacking the TATA box but in some cases may be replaced (as in *Drosophila* heat-shock genes) by a specific regulatory sequence such as the 15 nucleotide consensus CTNGAATNTTCTAGA between -60 and -80. This sequence is required for the positive activation of transcription in response to heat shock.

A different class of upstream regulatory elements that can stimulate transcription in an orientation-independent manner, and even work downstream of the gene, are the enhancers. These were first described as 72 bp repeats in SV40 DNA able to stimulate transcription of a transfected globin gene several hundred-fold, irrespective of their orientation. In polyoma virus, enhancers even determine the host range. They are a common feature of many viruses and some cellular genes, such as immunoglobulins, tyrosine aminotransferase and antithrombin III. Enhancers share a common or core sequence with the consensus TGTGGAATTAG.

As in yeast, *cis*-active sequences that repress transcription have also been found in higher eukaryote class II genes. An example is the tripartite site at the replication origin in SV40 which can bind large T-antigen and which thus represses its own synthesis. The ovalbumin gene also contains an enhancer near the TATA box and in addition a 'blocker' between positions -295 and -425. Deletion of this blocker sequence allows expression of the gene in oviduct tubular gland cells, where it is not normally active.

Introns occur in most class II genes, and are very variable both in number (up to 50 in collagen genes) and length (up to several thousand base pairs). The consensus sequence of boundaries between introns and exons is AGGTAAGT ... intron....TCNCAGG, with the GT and AG dinucleotides invariant at the 5' and 3' splice sites, respectively.

Termination of transcripts of class II genes remains poorly understood; present

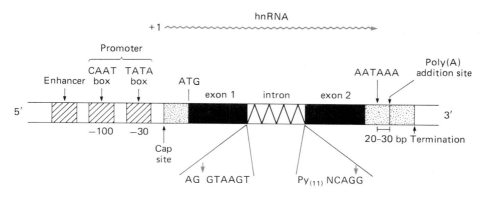

Figure 3.2. Class II gene structure. A typical gene is characterized by a promoter region at the 5'-end and the presence of introns. Enhancers are commonly found at the 5'-end of the gene although they may be internal or downstream of the gene. Transcription factor binding sites are also found at the 5'-end of these genes and in some cases the CAAT and TATA consensus sequences are absent. The left junction (splice donor) and right junction (splice acceptor) consensus sequences are shown together with the splice sites (↓).

evidence suggests it occurs hundreds or even thousands of nucleotides downstream of the 3'-end of mRNA, which in turn generally lies about 35 nucleotides downstream of the site coding for the polyadenylation signal AAUAAA. Recent evidence from SV40 indicates that AT-rich regions within GC-rich ones may be involved in termination by RNA polymerase II. A typical class II gene is shown in *Figure 3.2*.

1.4 Class III genes

These physically small genes are transcribed by RNA polymerase III. Most are reiterated and clustered; the oocyte-specific 5S genes of *Xenopus*, for example (see *Figure 3.3*), are present at more than 20 000 copies per haploid genome though most class III genes occur in tens or hundreds. The most striking feature of class III genes is the presence, clearly proven by deletion analyses, of internal promoters. In the *Xenopus* 5S gene the promoter lies between +45 and +83, whereas in tRNA genes it is split into two separate blocks (one between +8 and +30, the second from +51 to +72). Termination has been best studied in 5S genes, where it is caused by a run of four A residues set between two GC-rich regions. In higher eukaryotes, regions upstream of the genes are not required for transcription, though they may modulate its rate; but in lower eukaryotes, such as yeast and *Drosophila*, upstream sequences are needed for significant levels of transcription, as well as the internal promoter.

1.5 Chromatin structure

Another distinctive feature of eukaryotic DNA is its association with large amounts of specific proteins to form chromatin. In all eukaryotic cells, approximately 200 bp of DNA are associated with an octamer of histones to form

40 Gene structure and transcription

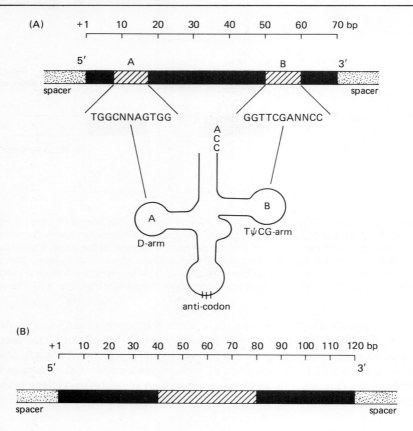

Figure 3.3. RNA polymerase III promoters. (**A**) tRNA gene showing the internal promoter consensus boxes A and B that correspond respectively to sequences in the D-arm and TψCG-arm of the mature tRNA. (**B**) The 5S gene has a single internal promoter sequence as indicated ▨. A transcriptional activator protein binds to the region between +47 and +96.

each repeating unit of structure, the nucleosome. Nucleosomes thus have two copies each of the four histones H2A, H2B, H3 and H4 with 140 bp of DNA around this core, plus a spacer region separating adjacent nucleosomes of between 10 and 100 bp. Histone H1 is present at one copy per nucleosome (though apparently absent in yeast), sealing the DNA entry/exit points and probably also complexing to adjacent nucleosomes. Nucleosomes are further wound into 30 nm fibres and higher order structures, ultimately producing a packaging ratio for the DNA of about 1000; that is the DNA is reduced to a thousandth of its original length. It is with this chromatin, rather than naked DNA, that factors regulating gene expression must interact.

The basic nucleosome structure of the chromatin is maintained during transcription, but there are differences between active genes and bulk (inactive) chromatin (*Figure 3.4*). In the case of the heat-shock genes of *Drosophila*, for

Gene structure and transcription machinery in eukaryotes

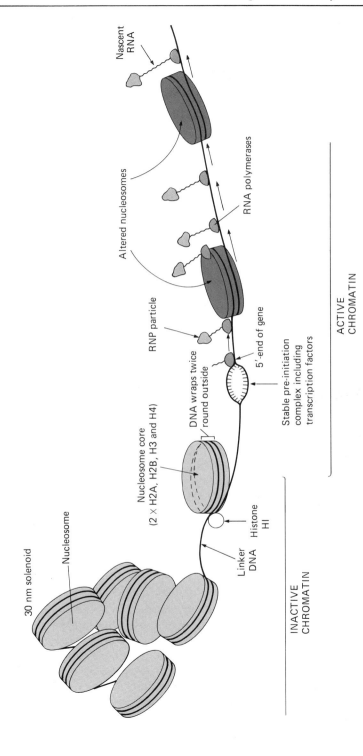

Figure 3.4. Structure of eukaryotic chromatin. The unwinding of inactive heterochromatin around an active gene, showing the persistence of nucleosomes in transcribed regions.

example, the nucleosomal repeat becomes much less precise following induction; and genes being transcribed or having the potential to be transcribed in a particular cell are especially sensitive to attack by DNase I. Furthermore, some sites are hypersensitive to this nuclease; many of these map in probable promoter regions (i.e. immediately upstream of the gene) and could be in important regulatory sequences (see Chapter 4). Genetic evidence from *Drosophila* shows that the nuclease hypersensitive sites of the *Sgs4* gene lie 350 bp upstream; deletions in this region abolish transcription *in vivo*. The structural basis of DNase sensitivity and hypersensitivity remains uncertain at present.

2. Transcriptional machinery in eukaryotes

2.1 RNA polymerase multiplicity

Analysis of solubilized RNA polymerases from eukaryotes yielded an exciting finding. Chromatography on DEAE–Sephadex revealed three activities eluting at different salt concentrations and demonstrating different sensitivities to an important inhibitor of RNA synthesis in eukaryotes, α-amanitin. These activities were annotated in order of elution as: RNA polymerase I (pol I), totally resistant to amanitin; RNA polymerase II (pol II), highly sensitive to the drug (inhibited by <0.1 μg/ml in higher eukaryotes); and RNA polymerase III (pol III), showing intermediate sensitivity and usually inhibited in the $10-500$ μg/ml range. (An alternative nomenclature used by some European groups is to call RNA polymerases I, II and III, polymerases A, B and C, respectively.) This pattern of RNA polymerases is maintained more or less unchanged throughout the eukaryota, although enzymes from lower eukaryotes are more variable in their response to α-amanitin (with generally less sensitivity). In yeast, for example, pol I is inhibited by high concentrations but pol III is totally resistant. We now know that pol I transcribes exclusively the large (nucleolar) rRNA genes, and generally no others; pol II is responsible for all pre-mRNA (hnRNA) transcription as well as that of many snRNAs, notably U1–U5; and pol III transcribes genes for tRNA, 5S rRNA and some other small RNA genes (such as U6 and some viral genes such as adenovirus VAI and VAII). It follows that *in vivo* pol I is essentially confined to the nucleolus, whereas the other two enzymes are nucleoplasmic.

2.2 Structure and function of nuclear RNA polymerases in eukaryotes

RNA polymerases I, II and III are each more complex than the single bacterial enzyme and are found in the nucleus of the cell. All polymerases are of high molecular weight ($>500\,000$) and contain multiple subunits, always including two with molecular weights greater than 100 000. Furthermore, there is often heterogeneity within each major class, though some of this complexity is artefactual, arising from partial proteolysis of subunits during purification. The best example of this is pol II, which in several organisms can be resolved into three species differing only in the sizes of their largest subunits. Thus in mammals, subunits IIo, IIa and IIb have apparent molecular weights of around 240 000, 215 000 and 180 000 daltons, respectively, with only one present in any particular polymerase molecule. It is now clear that these proteins are all the products of a single gene; almost certainly IIo is the major 'physiological' subunit, from which a C-terminal peptide can be cleaved to yield IIb; IIb is not seen by western blotting analysis of fresh tissue extracts with anti-pol II antibodies and enzyme containing IIb cannot initiate accurate transcription. The difference between IIo and IIa actually resides in the extent of phosphorylation of the C-terminal region absent from IIb and does not reflect a molecular weight difference at all; this unusual section of polypeptide contains 52 repeats of a consensus amino acid sequence (Tyr – Ser – Pro – Thr – Ser – Pro – Ser), in which the last serine is susceptible to phosphorylation. In *in vitro* transcription assays, RNA polymerase containing IIo (the phosphorylated form) is ten times more active than that with IIa on adenovirus promoters, but the physiological functions of IIo and IIa-type enzymes remain unclear at present.

To what extent are the three major RNA polymerases inter-related? The most thorough investigation of this question has been made with the yeast RNA polymerases, for which antibodies have been raised against every putative subunit of all three polymerases. In yeast, pol I (mol. wt 630 000) has 13 subunits, pol II (mol. wt 567 000) has 10 and pol III (mol. wt 697 000) has 14 (*Figure 3.5*). Within each enzyme class there is no evidence of homology between the various subunits and, furthermore, the two large subunits of all three enzymes seem to be unique gene products from six distinct loci. In the case of the smaller subunits things are different, however, and a number of these are common to two or even all three polymerases. Thus yeast pol I, II and III share three subunits of molecular weight 27 000, 23 000 and 14 500; and pol I and III share a further two subunits of molecular weight 40 000 and 19 000. This precise pattern is not universal, but the general principle of sharing some small subunits has been observed in plants and higher animals. In all, probably 25 different genetic loci are required to code for RNA polymerase components in yeast. These are found in single copies, and are widely scattered on different chromosomes. All the eukaryotic enzymes, like those of bacteria, contain two atoms of zinc. Recently, genes for several subunits of yeast RNA polymerases have been cloned and those for the largest subunits of pol II and III (and *Drosophila* pol II) have been partially sequenced; already substantial homologies have been noted between these genes and also with the gene coding for the β' subunit of the *E.coli* enzyme. Moreover, the second largest subunit of yeast pol II (mol. wt 138 750) has extensive homology with the β subunit of *E.coli* RNA polymerase, including the probable nucleotide- and zinc-binding regions.

2.3 Other eukaryotic RNA polymerases

Although the three types of enzyme discussed in the previous section constitute the overwhelming majority of RNA polymerase activity in eukaryotic cells, there are two other minority forms: one in mitochondria, and one in chloroplasts. The polypeptide components of the RNA polymerases of both of these organelles

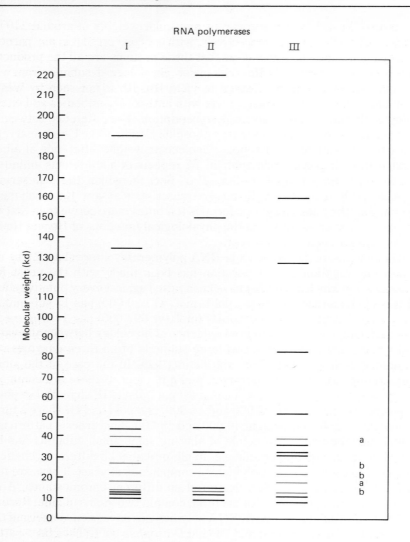

Figure 3.5. Subunit structures of yeast RNA polymerases. Subunit patterns of major RNA polymerase species emphasizing those subunits common to more than one enzyme. a, Common to RNA polymerases I and III; b, common to all three enzymes.

were until recently thought to be coded for by nuclear genes, and both enzymes are unaffected by the best characterized inhibitor of nuclear eukaryotic polymerases, α-amanitin. However, it is now known that at least some components of the chloroplast RNA polymerase are encoded within the chloroplast genome. Mitochondrial RNA polymerases are often single polypeptides and possibly the smallest RNA polymerases known (mol. wt <70 000), though some may be larger, and there is even evidence of multiplicity of RNA polymerase activities in yeast; perhaps related to the greater complexity of fungal mitochondrial DNA. Intriguingly, recent gene sequencing has revealed a striking homology between mitochondrial and T7 bacteriophage RNA polymerases. Chloroplast polymerases, on the other hand, are large and have a structure generally similar in terms of subunit composition to the bacterial enzyme. They are not inhibited by rifampicin but usually recognize multiple promoters of similar general arrangement to bacterial ones (i.e. with conserved promoter boxes at -10 and -35), though the consensus sequences in these promoters are distinct to chloroplast DNA.

2.4 Properties of nuclear eukaryotic RNA polymerases

Unlike bacterial RNA polymerase, none of the nuclear eukaryotic enzymes is able to initiate RNA synthesis from their cognate promoters when pure enzyme and DNA template are mixed *in vitro*. Despite the greater complexity of the eukaryotic polymerases, they are still deficient in their ability to initiate transcription accurately, and require extra factors to function correctly. Nevertheless, the three polymerases all synthesize RNA under so-called 'non-specific' assay conditions (i.e. when supplied with pure DNA, appropriate ionic environments and nucleoside triphosphate substrates). In non-specific assays the RNA polymerases initiate randomly from 'pseudopromoters', usually single-strand breaks ('nicks') in the DNA template. Pol II is especially bizarre, being almost inactive on relaxed native DNA from which nicks and single-stranded regions have been removed. However, it can form very stable complexes at nicks, and transcribes single-stranded DNA very actively (but not from true promoters). Transcriptional studies with chromatin templates have also been attempted in many laboratories; however, they have been plagued with difficulties that stem mainly from the ill-defined nature of the chromatin complex, and have yet to contribute significantly to work in this field.

3. Accurate *in vitro* transcription systems from eukaryotes

3.1 General features

Lack of success with purified enzymes led researchers to the use of cruder preparations in transcription assays. The most usual procedure is to incubate a cloned gene fragment (including the promoter) with a cell or nuclear lysate (after clarification by high-speed centrifugation) and in some cases supplemented with purified RNA polymerases. Using the so-called run-off assay, in which RNA is synthesized on a truncated template and analysed electrophoretically to detect species of RNA with the predicted length based on the known *in vivo* initiation site, fidelity has been demonstrated for all three polymerases. However, the template is in most cases not packaged into chromatin during these *in vitro* assays and so they are therefore substantially less than a perfect reconstruction of the situation *in vivo*.

The essential common feature to emerge from this work is that, unlike their prokaryotic counterparts, eukaryotic RNA polymerases recognize not a sequence of DNA but a pre-existing DNA – protein complex at gene promoters. Sequence identification is important for the correct initiation of transcription in eukaryotes, but in all cases is carried out by other protein(s) before the polymerase arrives. Details as to how this situation is achieved differ in detail between the three enzymes, and are only beginning to be understood.

3.2 Transcription by RNA polymerase I

What seems to happen in this case is that a species-specific protein factor binds at or near the promoter, followed by an 'activated' (possibly phosphorylated) form of pol I, to yield a pre-initiation complex. The generation of this complex *in vitro* is competitive with histone binding, but once established the factor – promoter complex is stable for many rounds of transcription and so can interact with many polymerases sequentially. Virtually nothing is known of the initiation and elongation reactions of pol I; but termination by this enzyme appears to be mediated by (as yet unknown) protein factors. In *Xenopus*, however, there is a change just after reading through the gene which causes transcription to continue but the RNA becomes unstable and subject to immediate degradation. The RNA polymerase traverses the spacer region in this mode, and is then available either for termination or for immediate initiation at the next promoter complex, presumably depending on the physiological needs of the cell at the time. It seems likely that a polymerase molecule could transcribe an indefinite series of tandem genes in this way, without ever leaving the template and becoming subject to the limitations of the diffusion time needed for location of the promoter.

3.3 Transcription by RNA polymerase II

There are now several cell-free systems in which accurate transcription by this enzyme is demonstrable, and many genes have been used in such lysates, including, for example, globins, ovalbumins, fibroins, protamines, histones and virus sequences such as SV40 and adenovirus. Recently accurate snRNA transcription (U1) has also been achieved *in vitro*.

As with pol I, a stable complex involving an as yet ill-defined number (probably at least three) of protein factors is first assembled at the core (TATA box) promoter and only then is the RNA polymerase molecule able to bind and initiate RNA synthesis. One of these factors apparently recognizes the TATA box sequence at position -25. Other separate classes of factors bind to upstream activator sequences; for example, a complex mixture of proteins with molecular weights in the range of 50 000 – 70 000 apparently recognizes the CAAT box. Careful use of the detergent sarkosyl (which inhibits different stages of complex formation depending upon the concentration employed) as well as kinetic studies have demonstrated an ordered assembly route for the formation of a transcription-competent complex. The initial binding of factors and polymerase (the latter requiring ATP hydrolysis) to the DNA, the 'commitment step', is followed by conversion to a 'rapid start complex', perhaps analogous to the open complex of prokaryotes. This conversion has a half-time of 9 min, after which initiation of transcription can begin immediately. The formation of just two phosphodiester bonds seems sufficient to commit the complex to elongation, a much lower value than the 10 or so needed to commit the bacterial enzyme. However, when given substrates that allow only one phosphodiester bond to form, pol II does enter an abortive initiation cycle similar to that observed with the *E.coli* polymerase. Furthermore, there is a tendency to 'stall' after 7 nucleotides have been polymerized and some further transition, to a fully stable elongation state, seems to occur at this point.

Elongation by pol II may also be facilitated by transcription factors, of which the best studied are the S-proteins. Antibodies against one of these, known as S-II, inhibit accurate initiation on the adenovirus promoter as well as elongation in isolated nuclei, and selectively stain nucleoplasmic chromatin in immunofluorescence studies. The mechanism of elongation is of particular interest in the context of chromatin structure, and it seems that pol II can somehow transcribe through nucleosomes without completely displacing histones from the DNA. There is preliminary evidence of termination factors for pol II, but little is yet known of this reaction.

3.4 Transcription by RNA polymerase III

Two proteins in *Xenopus*, TFIIIB and TFIIIC, are required for initiation on both 5S and tRNA genes while a third, TFIIIA, is needed only for the transcription of 5S genes. In this latter case TFIIIA is the first factor to recognize and bind to the internal promoter of the 5S gene; one molecule binds to each gene. Subsequently TFIIIC, then TFIIIB and finally RNA polymerase III also bind (*Figure 3.6*). Factor TFIIIC also has DNA-binding properties whereas TFIIIB probably interacts with proteins in the forming complex (it can also bind to pol III itself). ATP hydrolysis is involved in the generation of this very stable nucleoprotein complex, but other mechanistic aspects remain controversial. Factors TFIIIC and TFIIIB bind in the same order on tRNA genes, and among other things probably create a 'bend' in the DNA molecule. In this case the major initial interactions seem to be between factor TFIIIC and the 'B' block of the internal promoter. Once bound, these complexes are very stable and can support multiple (>40) rounds of transcription *in vitro*, with the RNA polymerase dissociating and reassociating for each event.

Termination of 5S transcription by pol III occurs at a distinct sequence in *Xenopus*, and requires no additional protein factors.

4. Processing of RNA in eukaryotes

Three major kinds of processing events have been identified in eukaryotic nuclear RNA. Firstly, class II gene transcripts are modified at their 5'-end by the addition of a so-called 'cap' structure, which plays a major role in translation. The cap is an added 5' terminal G, methylated on the 7-position of the base and linked to the initiator nucleotide by an unusual 5' – 5' triphosphate linkage (*Figure 3.7*).

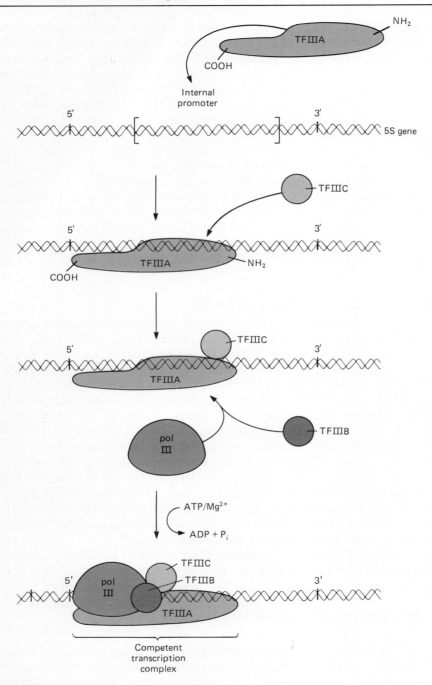

Figure 3.6. Transcription complex formation by RNA polymerase III. TFIIIA binds first to the internal 5S gene promoter followed by factor C (which may be composite), then B and finally RNA polymerase III itself to form the competent transcription complex.

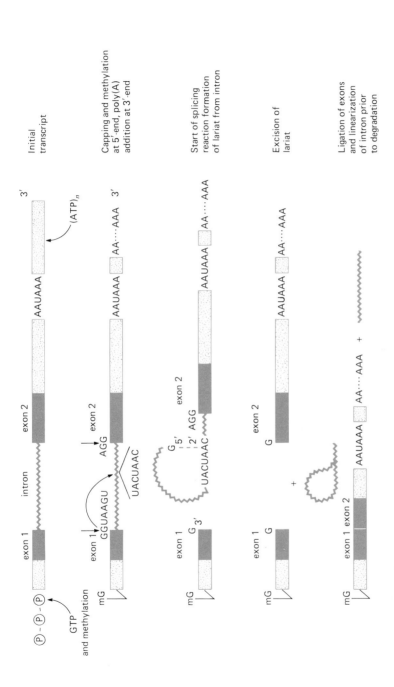

Figure 3.7. Stages in the processing of hnRNA to mRNA. The initial transcript is modified by capping at the 5'-end in a reaction involving GTP; the cap structure is then methylated. Polyadenylation occurs at the 3'-end of the transcript after the poly(A) addition signal AATAAA. The left end of the intron is cut (↓) and a lariat is formed between the G at the 5'-end of the intron, and the 3'-A nucleotide, a loose consensus sequence, TACTAAC, lies approximately 30–40 bases upstream of the 3'-end of the intron. The linkage is unusual in being a 5'–2' phosphodiester bond. The right end of the intron is cut, the lariat is released and the exons joined. The remaining lariat structure is degraded.

Secondly, the great majority of class II gene transcripts (with certain exceptions such as histone mRNAs) contain a 3'-poly(A) tail, 40–100 residues long, which is added post-transcriptionally. Addition of this tail, by a special polymerase, is triggered by the signal AAUAAA present in the 3'-non-coding region of the mRNA; this signal is sufficient by itself in frog β-globin mRNA, but the human equivalent requires extra GU- and U-rich sequences further downstream from it for maximal efficiency.

Finally, transcripts of all three gene classes are usually subject to endo- or exonucleolytic cleavage followed, in the case of introns in class II and some other genes, by exon splicing to form the functional mature species. 5S RNA is an exception, being synthesized directly in the mature form. Class I genes are transcribed as a single large precursor (45S, >13 kb in mammals) from which the mature 18S, 5.8S and 28S rRNAs are derived via a well-characterized cleavage pathway. In a class II gene transcript with multiple introns, these tend to be removed from the 5'-end of the RNA first; splicing proceeds by cleavage of any particular intron at its 5' (left) consensus sequence, the free end of the intron folds to form a lariat structure with the branch consensus box TACTAAC, itself located in the intron near the 3'-end splice consensus. The G residue at the 5'-end of the intron at the 5'-end splice junction then reacts with the 2'-hydroxyl position on the ribose of the A residue at the 3'-end of the RNA sequence UACUAAC. The 3' slice site is then cleaved, the two exons ligated together to form a continuous coding sequence and the excised intron is linearized (*Figure 3.7*). This ATP-requiring reaction is probably carried out by a 50–60S 'spliceosome' complex containing proteins and snRNAs, such as U1 which has sequence complementarity to the 5' exon–intron splice site (GUAAGUA), and U2 which has sequences complementary to the consensus sequence at the branch site.

This form of intramolecular intron removal and exon splicing is undoubtedly the most common pathway in eukaryotic nuclei. Evidence is also mounting that alternative splicing provides an important means of making multiple transcripts from a single gene often in a tissue-specific manner (see Chapter 4, Section 3.3). However, there is also evidence from trypanosomes and nematodes which suggests that *trans* splicing can also occur. In trypanosomes a common 35 nucleotide leader sequence is added to virtually all mRNAs post-transcriptionally, and in nematodes a common leader is spliced onto the 5'-end of three of the four actin mRNAs. The significance and indeed the extent of such *trans* splicing remains unclear.

An intriguing recent discovery is that primary transcripts can, in some cases, process themselves. Self-splicing *in vitro* is a property of 26S rRNA from *Tetrahymena* and of some mitochondrial DNA transcripts from lower eukaryotes; intron removal requires only a guanine nucleotide, magnesium ions and monovalent cations. These RNA molecules are truly autocatalytic, requiring no proteins to assist in the splicing reaction, and have been termed 'ribozymes'.

5. Further reading

Baker,S.M. and Platt,T. (1986) Pol I transcription: which comes first, the end or the beginning? *Cell*, **47**, 839.
Birnstiel,M.L., Busslinger,M. and Strub,K. (1985) Transcription termination and 3′ processing: the end is in sight. *Cell*, **41**, 349.
Breitbart,R.E., Andreadis,A. and Nadal-Ginard,B. (1987) Alternative splicing: a ubiquitous mechanism for the generation of multiple protein isoforms from single genes. *Annu. Rev. Biochem.*, **56**, 467.
Dynan,W.S. (1986) Promoters for housekeeping genes. *Trends Genet.*, **2**, 196.
Dynan,W.S. and Tijan,R. (1985) Control of eukaryotic mRNA synthesis by sequence specific DNA binding proteins. *Nature*, **316**, 774.
Grosveld,F., Blom van Assendelft, Grjaves,D.R. and Kollias,G. (1987) Position-independent, high-level expression of the human β-globin gene in transgenic mice. *Cell*, **51**, 975.
Maniatis,T. and Read,R. (1987) The role of small nuclear ribonucleoprotein particles in pre-mRNA splicing. *Nature*, **325**, 673.
Meytrowitz,E.M., Raghavan,K.V., Mathers,P.H. and Roark,M. (1987) How *Drosophila* larvae make glue: control of *Sgs3* gene expression. *Trends Genet.*, **3**, 288.
Sentenac,A. (1985) Eukaryotic RNA polymerases. *Crit. Rev. Biochem.*, **18**, 31.

Regulation of transcription in eukaryotes

1. Introduction

Transcriptional regulation in eukaryotes can occur at a variety of levels. As with prokaryotes, proteins can interact with DNA sequences to either stimulate or repress activity of a particular gene; but quite unlike prokaryotes, changes may also occur via differential usage of promoter, termination and splice sequences in various tissues to produce different mRNAs from the same gene.

2. Regulation of class I rRNA genes in eukaryotes

2.1 General aspects

There is a good reason to believe that rRNA synthesis is co-ordinately regulated in eukaryotic cells. Different rates of rRNA synthesis have been demonstrable *in vivo* under various circumstances which mainly relate, in the ultimate analysis, to the growth and proliferation state of the cell. Thus cells of regenerating liver synthesize rRNA at some ten-fold greater rates than normal hepatocytes; tumour cells, proliferating lymphocytes and early embryonic cells are likewise very active in terms of their rates of rRNA synthesis. Increased rRNA production is not always mediated at the transcriptional level, but probably is when major or long-term changes are involved. There is also a kind of eukaryotic 'stringent response' in the sense that ribosome production and rates of protein synthesis are inter-dependent in most cells but there does not seem to be any universal mechanism directly comparable with that seen in bacteria. No novel nucleotides comparable with ppGpp (see Chapter 2) have been implicated in eukaryotes, and depression of rRNA production following starvation may, at least in some cells, reflect post-transcriptional processing and control at the level of the rate of degradation (a phenomenon known as 'wastage'). Wastage of rRNA can be quite significant in cells but is usually much reduced when the growth of cells is stimulated. Examination of RNA polymerase 'levels' under these conditions of variable rRNA synthesis, usually after solubilization and by the use of non-specific assays, has led to the general conclusion that polymerase activity, probably reflecting numbers of active molecules, does not usually show a simple relationship to rates of rRNA synthesis *in vivo*. A good example of this is seen during oogenesis in *Xenopus*. Temporal regulation of 5S rRNA, hnRNA (lampbrush activity) and large (40S) rRNA transcription during oocyte development has been thoroughly characterized, and each of these classes of gene exhibits specific peaks of activity as the cell grows. RNA polymerase content, however, increases progressively (and overall by a massive 10^4 to 10^5-fold) with respect to all three species and not at all in concert with the peaks of activity of the genes that they are known to transcribe. However, in rapidly growing or dividing cells, exhibiting high levels of rRNA synthesis, there does seem to be both a relative (to pol II) and absolute increase in the amounts of pol I and pol III when compared with resting or non-proliferating cells. Phytohaemagglutinin-stimulated lymphocytes, for example have 17 times more pol I and pol III activity but only eight times more pol II activity than their unstimulated counterparts though these levels are reached long after the major increase in transcription rates (*Figure 4.1*). It has been calculated that mouse myeloma cells, which have very high rates of rRNA synthesis, probably each contain about 2×10^5 molecules of pol I, which, in turn, are sufficient to saturate the whole complement of the cell's reiterated ribosomal genes.

This experimental approach is clearly insufficient to explain the mechanisms of ribosomal gene regulation. Is the implicit assumption that cells can vary the

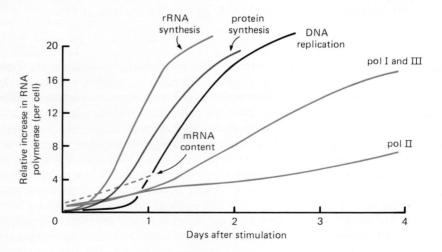

Figure 4.1. Events following mitogenic stimulation of lymphocytes. Rates of macromolecule synthesis following lymphocyte activation. Note that increases in polymerase levels are late events relative to general changes in gene expression and replication.

levels of their pol I and pol III proteins correct? If so, it is unlikely to be the only process at work. Covalent modification of RNA polymerases, notably phosphorylation, can also occur though its significance remains obscure. In yeast several subunits of all three enzymes are phosphorylated, to varying extents, *in vivo*; but in animal and plant cells only pol II seems to be modified in this way, and indeed this is the most active form of the enzyme. It now seems likely that a third aspect, the abundance and activities of transcription factors, also has widespread importance in the regulation of transcription rates.

2.2 Encystment in Acanthamoeba

Although other organisms (especially mouse and human) have provided useful insights into the factors involved in ribosomal gene transcription, progress on the regulation of these genes is probably most advanced in the case of the protozoan *Acanthamoeba castellanii*. This unicellular aquatic organism undergoes encystment when environmental conditions become adverse, a process accompanied by a general shutdown of many genes and in particular a dramatic down-regulation of rRNA synthesis.

As mentioned earlier, at least two elements are involved in promoter recognition and initiation by RNA polymerase I. One is a highly species-specific protein (mol. wt >100 000), and is involved in promoter-binding. In *Acanthamoeba* it interacts by itself with DNA between positions -14 and -67. After association of the polymerase with DNA, the protein–DNA interactions are extended downstream to position $+18$. The second 'factor' seems to be an activated form of the polymerase itself, the availability of which apparently limits the overall rate of rRNA transcription.

The large ribosomal genes of *Acanthamoeba* code for a primary transcript of 39S, which is later processed to the functional ribosomal species. When cells are shifted from a nutrient-rich to an impoverished medium, dormant cyst formation is triggered and this is accompanied by a sharp decrease in the synthesis of 39S rRNA. This decrease reaches zero within about 8 h, during which period the amounts of pol I and of the promoter-binding transcription factor do not change in any detectable way. Nevertheless, extracts from *Acanthamoeba* cells isolated at various times during encystment show a progressive loss of ability to initiate rRNA transcription *in vitro*. This change seems to involve a stable alteration of the polymerase; it can be reversed by supplementation of pol I from vegetative cells. Thus although non-specific assays reveal no differences in pol I activity during encystment, subtle but important changes are evidently occurring. These changes are reflected in much greater thermolability of the RNA pol I of the cyst, but no changes in subunit composition have been detected.

3. Regulation of class II genes

3.1 Transcription factors in lower eukaryotes

The best understood examples in this area come from studies with yeast. The yeast protein HAP1 binds to the upstream activation site (UAS1) of the yeast iso-1-cytochrome *c* (*CYC1*) gene, in some way stimulated by haem, both *in vitro* and *in vivo*; the consequence *in vivo* is stimulation of transcription. Since haem is synthesized in mitochondria, it must migrate from this site to the nucleus and thus mitochondrial metabolism influences nuclear gene activity. It has recently been found that a second protein, RC2, binds to the same site in UAS1 as does HAP1. RC2 requires haem for its synthesis or stability, but not for DNA-binding, and may act as a negative regulator of *CYC1*.

The *GAL4* protein, 881 amino acids in size, positively regulates at least five structural genes including *GAL1*, the product of which is involved in galactose metabolism. *GAL4* binds to a UAS. If the 73 amino acids of the N-terminal, DNA-binding domain are replaced by the DNA-binding domain of an *Escherichia coli* protein, *lexA*, then this fusion protein fails to activate transcription of the *GAL* genes. However, if the normal UAS that binds to *GAL4* is also replaced by the *lexA* operator then transcriptional stimulation is seen with the fusion protein, even if this operator is placed in an intron downstream of the site of initiation of transcription. This proves convincingly that it is the C-terminal part of *GAL4* that causes transcriptional activation, not some consequence of a specific DNA–protein interaction involving the secondary structure of DNA. As a further control, *lexA*-binding does not stimulate transcription from the *lexA* operator when this is put into yeast.

The *GAL* genes are also under negative control; protein *GAL80* blocks transcriptional activation by *GAL4*, by binding directly with *GAL4* attached to the UAS. This happens in the absence of galactose, but it is proposed that the presence of the inducer (galactose?) causes *GAL80* to dissociate and permits a transcription factor to take its place. Consistent with this model are the observations that the 30 amino acids at the C-terminal of *GAL4* are needed for the interaction with *GAL80*, and that *GAL4* remains bound to the UAS in the absence or presence of transcription (*Figure 4.2*).

The generality of this kind of control in eukaryotes is implied by other fusion experiments. When the DNA-binding region of yeast protein *GLN4* (a positive regulator of several genes involved in amino acid biosynthesis) is replaced with the DNA-binding domain of the chicken *jun* oncoprotein (which causes sarcomas), the fusion protein binds to the same sites in yeast DNA as *GLN4*. This suggests that *jun* is a DNA-binding protein which may activate other cellular genes in the chicken.

The tight linkage between transcription and translation has facilitated the use of attenuation as a transcriptional regulation mechanism in prokaryotes (see Chapter 2). It is therefore surprising that in yeast, where these processes are physically disparate, attenuation control also occurs. Deletion of parts of the leader of *GCN4* mRNA increases the efficiency of translation; regulatory proteins normally bind to this leader and inhibit translation, thus restricting the amount of *GCN4* available to bind to the UASs of amino acid biosynthesis genes.

3.2 Transcription factors in higher eukaryotes

Affinity chromatography with specific DNA sequences, DNA footprinting and molecular genetic analyses have identified several putative regulators in higher

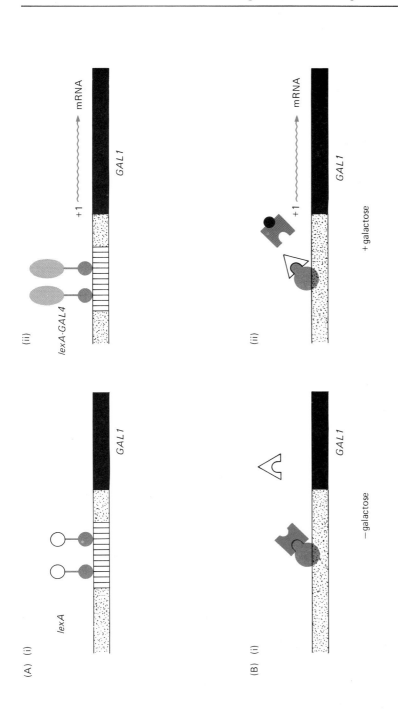

Figure 4.2. (A) Activation of transcription of *lexA – GAL4* fusion protein. The bacterial *lexA* operator (▭) is inserted upstream of the *GAL1* gene. *lexA* produced in yeast will bind the operator but will not stimulate transcription (i). However, a fusion protein made up of the *lexA* DNA binding domain at the N-terminal, and *GAL4* sequences at the C-terminal will stimulate transcription (ii). (B) Negative regulation of the *GAL1* gene. (i) In the absence of the inducer (galactose) the *GAL80* protein (▼) binds to the *GAL4* protein(●) preventing interaction with transcription factor(s) (▲). (ii) In the presence of galactose (●) the system is induced by galactose binding to the *GAL80* protein causing it to dissociate from *GAL4* thus allowing the transcription factor(s) to bind and allowing the synthesis of *GAL1* mRNA.

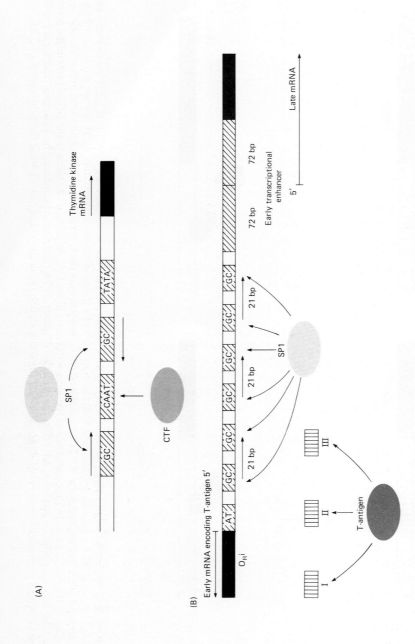

Figure 4.3. Transcription factor binding domains of two promoter regions. (**A**) Herpes simplex thymidine kinase gene. SP1 binds to GC-rich regions either side of the CAAT box, which itself binds the CAAT transcription factor (CTF). (**B**) SV40 has SP1 binding sites, as well as three T-antigen binding domains by which T-antigen regulates transcription. T-antigen autoregulates by binding strongly to sites I and II and weakly to III thus repressing early transcription. Studies on SV40 show that DNA-binding proteins in mammals can both stimulate and repress transcription.

eukaryotes and their viruses. SV40 T-antigen has already been mentioned (Chapter 3); also in the SV40 regulatory region is a series of 21 bp repeats, each containing two copies of the sequence GGGCGG; these sequences bind the well-characterized transcription factor SP1. Similar SP1-binding regions have been found in the Herpes simplex I thymidine kinase (TK) gene, in opposite orientations each side of the CAAT sequence (*Figure 4.3*). A factor binding to this region has also been identified. Some genes lacking a TATA box (such as hypoxanthine phosphoribosyltransferase, HPRT) have multiple SP1 binding sites and are probably under SP1 control. It may be that methylation of C residues in these motifs is involved in reduced SP1-binding thus decreasing the level of gene transcription.

More and more factors binding to UASs are being discovered in higher eukaryotes. These include the glucocorticoid receptor and AP1, a protein that binds to enhancers of genes activated by the tumour promoter TPA. TPA may modify AP1 via its known stimulatory effect on protein kinase C. In *Drosophila*, a heat-shock transcription factor (HSTF) binds to the heat-shock consensus element (HSE), following which some of the DNase I hypersensitive sites in it become resistant to nuclease digestion, presumably as a result of protection by HSTF.

The five human β-like globin genes are arranged as a cluster $5'\epsilon\text{-}G_\gamma\text{-}A_\gamma\text{-}\delta\text{-}\beta\ 3'$ in a 60 kb region on chromosome 11. These genes are regulated in a tissue-specific and developmental-specific manner. The eukaryotic ϵ gene is expressed in the yolk sac, the G_γ and A_γ are expressed in the fetal liver, and the δ and β genes are expressed in the bone marrow.

Four elements have been identified that regulate β-globin gene expression. There is a positively activating promoter sequence as well as a putative negatively acting sequence in the promoter region. Also there are two downstream enhancers, one within the β-globin gene and one 800 bp downstream of the β-globin gene. This latter element is subject to developmental stage-specific regulation.

Another level of control also exists. Some β-thalassaemia patients have large deletions that remove DNA 100 kb upstream of the β-globin gene, leaving all the regulatory sequences described above, yet the β-globin gene is not transcribed. The deletion must therefore remove a control sequence that overrides the regulatory sequences immediately flanking the β-globin gene. A likely possibility is that this region controls chromatin structure. A candidate for these regulators are the DNase I superhypersensitive sites found upstream of the ϵ-globin gene and downstream of the β-globin gene. These sites are specific for erythroid cells and are present when any of the genes in the locus are being expressed. Evidence that these DNase I superhypersensitive sequences are responsible for regulating expression comes from the construction of transgenic mice with the β-globin gene. In transgenic mice the human β-globin gene is not normally expressed in the correct tissues, (e.g. expression has been reported in the testes) and expression (if any) is always at a low level. This presumably is the result of random integration of the human globin gene into the mouse genome. However, position-independent and tissue-specific expression can be obtained by making human β-globin constructs that contain the potential super-hypersensitive sites. These sites, which may themselves be enhancer elements, or sequences that bind to the nuclear matrix are dominant control regions that set up the β-globin complex for regulation by *trans*-acting factors. In non-erythroid cells of transgenic mice the sites are not superhypersensitive and the β-globin genes are not expressed.

Evidently there are similarities of principle with prokaryotes, in that genes may be up- or down-regulated by protein factors acting in *trans*, via protein–protein interactions. There are of course substantial differences in detail, partly consequent upon altered gene organization (e.g. lack of operons and the packaging of the DNA into chromatin) but the most telling differences occur as a result of multiple promoter usage and variable RNA-processing pathways.

3.3 Alternative promoters, splicing and polyadenylation

There are now several examples of genes where the transcription start sites differ between tissues. They include *Drosophila* actin, alcohol dehydrogenase and *antennapedia*; mouse α-amylase, Thy-1 and myosin light chain. The general result is different 5' exons on mRNAs in various tissues in which the genes are expressed. An example of tissue-specific alternative polyadenylation is found in the human gene encoding calcitonin. Use of a polyadenylation site in the fourth exon yields calcitonin mRNA in thyroid cells, whereas use of a site in the sixth (final) exon in brain results in a mRNA for a calcitonin-related peptide. Alternative polyadenylation signals in immunoglobulin genes have striking consequences, determining whether the proteins are secreted or membrane-bound.

Some genes are spliced differently in a tissue-specific manner despite being transcribed from the same promoter and utilizing the same polyadenylation signal. A spectacular example is the *Drosophila* P element. In the germ line only, a specific intron is spliced out yielding a mRNA for a transposase that causes the P element to move around. This splicing does not occur in somatic cells, and the transposase is not produced in them.

To date more than 50 examples have been found of genes regulated by alternative transcript processing pathways (*Table 4.1*); perhaps the most complex is rat tropomyosin, which has seven such pathways (*Figure 4.4*). In most cases alternative splicing produces variant proteins, or mRNAs translated with different efficiencies, but in *Drosophila* the *Dunce* gene contains an intron which itself codes for two other proteins, one being the gene *Sgs4*. The mechanisms which effect these differential processing routes remain totally obscure.

Table 4.1. Examples of cellular genes that show alternative splicing

Gene	Organism	Splice pattern	Regulation
Alcohol dehydrogenase	*Drosophila*	Alternative 5' exon	Developmental
Calcitonin	Human	Alternative 3' exon	Tissue-specific
Nerve growth factor	Mouse	Cassette	Tissue-specific
P-element	*Drosophila*	Retained intron	Tissue-specific

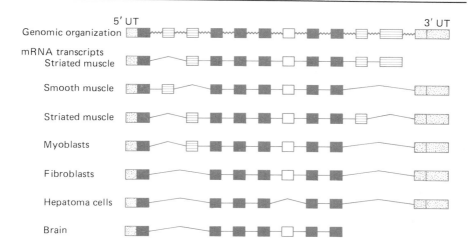

Figure 4.4. Alternative splicing of the rat α-tropomyosin gene. The rat α-tropomyosin gene is capable of following at least seven different splicing pathways. The organization of the exons (full, white and lined boxes), introns (zigzags) and untranslated (UT) regions (dotted boxes) in the genomic DNA is shown. The exons common to all tissues are shown as full boxes, muscle-specific exons as lined boxes and variable exons as white boxes.

4. Regulation of class III genes

4.1 Oogenesis and embryogenesis in Xenopus

There are two major types of 5S genes in *Xenopus*, the sequences of which differ in only six positions out of the 120 bp of the coding region. Both are reiterated and clustered, but to very different extents: the somatic 5S genes are present in just a few hundred copies per haploid genome and are found at the end (telomeric region) of a single chromosome. Oocyte 5S genes, on the other hand, are present at about 23 000 copies per haploid genome and occur clustered on multiple telomeres. As their name suggests, oocyte 5S genes are expressed during oogenesis and more or less only then; transcription becomes virtually undetectable early during embryogenesis and remains so in all somatic cell types. Oocyte 5S gene expression is maximal in very young (small) oocytes where it constitutes the great majority of all the RNA being synthesized, but declines in both relative and absolute terms as the oocyte grows and matures. Somatic 5S genes are transcribed in oocytes (though contributing <10% of total 5S synthesis) but remain active throughout the life of somatic cells. 5S RNA synthesis early in oogenesis is not co-ordinated with production of the larger rRNAs, which occurs later when the 40S nucleolar genes are activated; the 5S RNA therefore requires storage and in small oocytes this is accomplished by sequestration into 7S and 42S ribonucleoprotein particles. Ultimately all the rRNAs are used to generate the enormous ($\sim 10^{12}$) ribosome complement of the mature oocyte needed to sustain rapid rates of protein synthesis during early embryogenesis. The fully grown oocyte and the early embryo (pre-blastula stages) are essentially dormant with respect to most types of RNA transcription, including 5S species; this resumes during blastulation, but oocyte 5S expression is transient and has ceased by the middle of this stage.

There is therefore a well-characterized developmental regulation as 5S gene transcription in *Xenopus* (see *Figure 4.5*), though the functional differences between oocyte and somatic 5S RNA in *Xenopus* remain quite unknown.

So how are these switches in 5S gene transcription brought about? As outlined above, changes in the overall levels of RNA polymerase III cannot be responsible; the enzyme is in fact accumulating during a general downward trend in 5S gene transcription. One key to understanding the regulation of transcription of 5S genes has come from work with the specific transcription factor TFIIIA. This protein has two properties that are crucial to the regulation of 5S genes. Firstly, it has a higher affinity (about 4-fold) for somatic rather than oocyte genes. Secondly, it also binds tightly to 5S RNA and constitutes the protein moiety of the 7S storage particle. Early in oogenesis, when 5S transcription is very active, TFIIIA levels are also very high ($\sim 10^{12}$ molecules/oocyte) and constitute as much as 10% of the total cellular protein. This declines together with the rate of synthesis of 5S RNA to less than 10^{10} molecules per cell during ovulation, and rapidly falls below 10^4 per cell (the typical somatic value) after the multiple divisions of early embryogenesis. These changes are mirrored in the levels of TFIIIA mRNA over the same developmental period; early oocytes contain about 5×10^6 molecules of TFIIIA mRNA, but this falls to around 1×10^6 molecules at maturation and dramatically to only about one per somatic cell at the swimming tadpole stage.

Evidently there is a good correlation between TFIIIA gene expression and total 5S gene transcription rates, but how is the change from oocyte to somatic gene dominance brought about? Early in oogenesis there are more than enough TFIIIA molecules to activate all of the oocyte and somatic genes (since only one factor molecule binds per gene) and this is what occurs in the first instance. Accumulation of 5S RNA then becomes autoregulatory, since TFIIIA binds with the transcript to form the 7S storage particle; factor sequestered in this way is unavailable to mediate further transcription, and since mRNA levels (and hence TFIIIA production) is also declining, a decrease in 5S RNA synthesis becomes inevitable. At this point the selectivity of TFIIIA in favour of somatic genes becomes important; as factor levels decline, oocyte genes turn off preferentially to somatic ones. An 'all-or-nothing' situation seems to exist such that below a factor:gene ratio of 1:1, no genes will form active complexes. The mechanism for this remains uncertain, but may derive from a rapid exchange of TFIIIA between 5S gene sequences that can occur indefinitely, preventing stable complex formation, when free DNA is in excess. Only when all sites have TFIIIA bound (i.e. when factor is in excess) does exchange stop and a potential initiation complex forms (requiring subsequent addition of TFIIIC, TFIIIB and RNA pol III, see Chapter 2).

Formation of these initiation complexes is also competitive with repression by histone binding and conversion into inactive chromatin. In particular, the

Regulation of transcription in eukaryotes

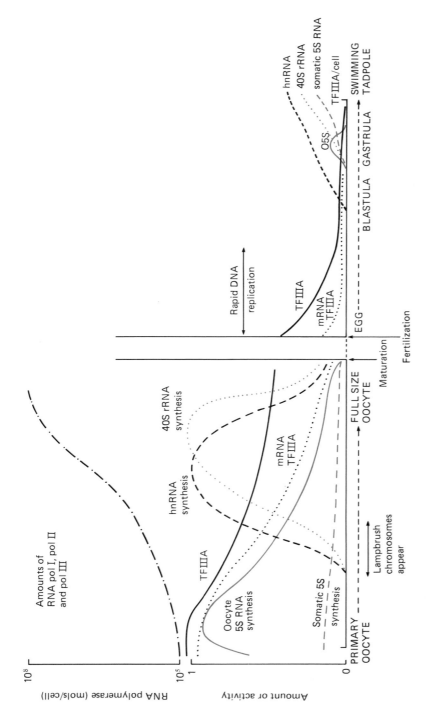

Figure 4.5. Patterns of transcription during oogenesis and embryogenesis in *Xenopus laevis*.

binding of histone H1 prevents active complex formation and removal of H1 allows binding to occur. The choice of commitment to activity or repression is therefore faced by 5S (and probably other) genes during every cell cycle when replication of DNA occurs. These features conspire to promote somatic rather than oocyte 5S RNA synthesis from embryogenesis onwards: TFIIIA levels fall lower and lower over this phase of development and come to constitute the limiting component in the formation of active complexes. So during early embryogenesis, when 5S genes replicate, only somatic ones are effectively converted to transcriptionally competent forms (because, due to their much smaller numbers, the ratio of TFIIIA to gene remains >1). Somatic 5S genes also replicate earlier than oocyte genes, compounding the situation to ensure that when the oocyte genes do divide there is insufficient factor to activate any of them.

Recently several studies have indicated that the regulation of 5S genes is more complex than originally thought. In particular, sequences outside the TFIIIA-binding region strongly influence the relative efficiencies of somatic and oocyte genes, and in some cell extracts TFIIIC, rather than TFIIIA, seems to be the limiting factor for transcription of the 5S genes.

Both the TFIIIA cDNA and the gene itself have been cloned and fully sequenced. The gene, of which there is just one copy per haploid genome, contains a typical pol II promoter, with TATA and CAAT boxes centred at positions -32 and -96, respectively. It spans altogether some 11 kb of DNA, with nine exons separated by eight introns. There is also the usual polyadenylation signal 255 bp downstream of the translation stop codon, altogether a very standard-looking structural gene. However, there are some striking features (apart from those relating to the structure of the protein) which may repay investigation; two stretches of $8-11$ nucleotides in the functional mRNA are highly homologous with tRNA 'B-box' internal promoter sequences, and another stretch of more than 40 nucleotides contains homologies with the TFIIIA-binding region of the internal promoter of the 5S gene. The implications of these observations for the autoregulation of TFIIIA transcription have yet to be explored, and it will now be necessary to elucidate how the regulation of a pol III type of gene might occur in a polymerase II transcription unit.

4.2 How does TFIIIA work?

TFIIIA has turned out to be a very interesting protein, with features quite different from those of classical prokaryotic DNA-binding proteins (notably repressors, see Chapter 2). TFIIIA is highly asymmetric in its three-dimensional conformation and consists of 344 amino acids which constitute two distinct types of domain. The first is a repeating unit of approximately 30 amino acids, present in nine tandemly linked copies, at the N-terminal end of the protein while the second is a quite different series of about 70 amino acids at the C-terminal end of the molecule. This latter domain is not involved in DNA-binding but is essential for transcription *in vitro*; its activity is orientated towards the 5'-end of the internal promoter, and it is thought to interact with other protein factors or with pol III itself. It is the nine repeating units which play the crucial role in promoter

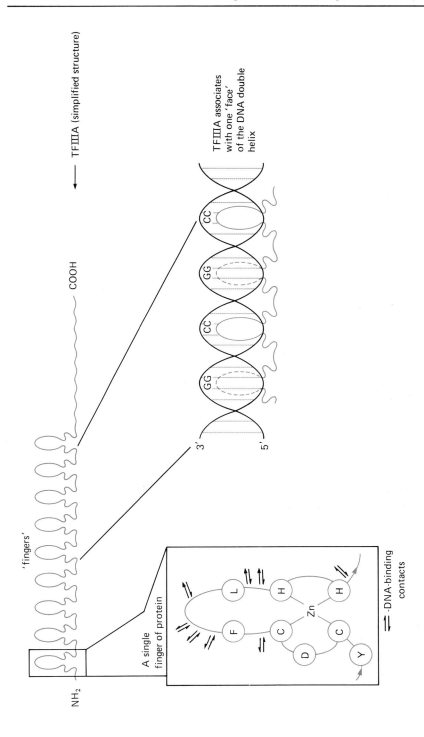

Figure 4.6. Interaction of TFIIIA and 5S RNA. The nine zinc-containing 'fingers' of TFIIIA are clustered at the N-terminal end of the protein. These fingers probably interact with one 'face' of the DNA double helix.

Table 4.2. Putative or definite 'zinc finger' proteins

Species	Protein
E.coli	Gene 32 protein
	UvrA
Yeast	ADR1
	SW15
	GAL4
	PPR1
	ARGRII
Drosophila	Serendipity
	Kruppel
	Hunchback
Xenopus	TFIIIA
	Xfin
Chicken	Oestrogen receptor
	Progesterone receptor
Mouse	mk1
	mk2
Rat	Glucocorticoid receptor
Human	Oestrogen receptor
	Glucocorticoid receptor
	C-erb A (thyroid hormone receptor)
	SP1

recognition and binding; each is associated with a zinc atom and loops outwards from it, with the result that TFIIIA has been described as a 'finger protein'. These nine fingers are linked by flexible joints, with the hydrophobic core regions (constituted by phenylalanine and leucine side chains) and positively charged outer groups capable of forming ionic bonds with the DNA phosphate backbone (*Figure 4.6*).

TFIIIA interacts with about 50 bp of DNA (between residues +45 and +97) within the 5S gene. Recent studies have indicated just how this binding probably occurs. Protection studies with nucleases and chemicals indicate multiple (nine) contact points about 5.5 bp apart and spanning the whole of the binding region. There is also a sequence periodicity, notably a run of one to three G residues, which repeats in the 5S promoter at about 5.5 bp intervals. The suggestion is that fingers of the TFIIIA protein loop into the DNA major grooves to make these contacts with the 'half-turn' repeats in the DNA.

The finger-contact model may be important in understanding a puzzling problem about pol III transcription; how can the enzyme read through a gene to which other protein factors are necessarily bound? An obvious possibility now is that the TFIIIA fingers are displaced and re-associate sequentially as the polymerase passes by, so that the factor does not at any time need to detach completely from the DNA. The fact that the factor binds most intimately to eight residues on the non-coding strand and only one on the coding strand is consistent with such a model.

It is increasingly clear that TFIIIA is not a special case, but that this kind of protein–DNA binding may be quite widespread in eukaryotes. Evidence of this comes from finding similar repeating units in other proteins and in other genes; several *Drosophila* gene products (such as *Serendipity*, beta, delta and Kruppel) and at least one from yeast (*ADR1*) are probably finger proteins. Not only that, but other types of gene promoters show 5'-nucleotide sequence periodicities that may infer binding to finger proteins; this is true of tRNA 'A' and 'B' boxes and two analogous boxes in *Alu* genes. More surprisingly, something similar may occur in regions controlling pol II genes, such as the SP1 binding zone in SV40. *Table 4.2* is a partial list of proteins thought to contain DNA-binding 'zinc fingers'. Indeed, recent studies with snRNA U6 which reveal sequences normally associated with the binding of pol II and pol III transcription factors on the same gene (which is apparently transcribed by pol III *in vivo*) are beginning to indicate a very complex interplay between the different classes of eukaryotic genes and the mechanisms which regulate them. Certainly there is, in the eukaryotic DNA-binding proteins investigated so far, a substantial difference in principle from the helix–turn–helix structures of their prokaryotic counterparts.

5. A rapidly changing scene

Some recent exciting developments indicate that, once again, certain cherished ideas will need revision. The divisions between class I, II and III genes are looking increasingly indistinct since the discoveries that variable surface glycoprotein genes in *Trypanosoma brucei* are transcribed by RNA polymerase I, probably after physical translocation to the nucleolus; and that snRNA U6 has type II and type III promoter elements, is transcribed by pol III but is only dependent upon conservation of the type II promoter sequences! Furthermore, it has now been found that the *lac* operator–repressor system can function in eukaryotic cells to regulate eukaryotic genes, after suitable fusion and transfection. Perhaps prokaryotes and eukaryotes are not so different after all...

6. Further reading

Bird,A. (1987) CpG islands as gene markers in the vertebrate nucleus. *Trends Genet.*, **3**, 342.
Breitbart,R.E., Andreadis,A. and Nadal-Ginard,B. (1987) Alternative splicing: a ubiquitous mechanism for the generation of multiple protein isoforms from single genes. *Annu. Rev. Biochem.*, **56**, 467.
Brent,R. (1985) Repression of transcription in yeast. *Cell*, **42**, 3.
Davies,K.E. and Read,A. (1988) *Molecular Basis of Inherited Diseases* (In Focus series). IRL Press, Oxford.
Harrison,S.C. (1986) Fingers and DNA half-turns. *Nature*, **322**, 597.
Hu,M.C.T. and Davidson,N. (1987) The inducible *lac* operator–repressor system is functional in mammalian cells. *Cell*, **48**, 555.

Johnson,S.A., Salmeron,J.M. and Pincher,S. (1987) Interaction of positive and negative regulatory proteins in the galactose regulation of yeast. *Cell*, **50**, 143.

McKnight,S. and Tijan,R. (1986) Transcriptional selectivity of viral genes in mammalian cells. *Cell*, **46**, 795.

Pawson,T. (1987) Transcription factors as oncogenes. *Trends Genet.*, **3**, 333.

Sassone-Corsi,P. and Borrelli,E. (1986) Transcriptional regulation by *trans*-acting factors. *Trends Genet.*, **2**, 215.

Struhl,K. (1987) Promoters, activator proteins and the mechanism of transcriptional initiation in yeast. *Cell*, **49**, 295.

Taylor,W., Jackson,I.J., Siegel,N., Kumar,A. and Brown,D.D. (1986) The developmental expression of the gene for TFIIIA in *Xenopus laevis*. *Nucleic Acids Res.*, **14**, 6185.

Tower,J. and Sollner-Webb,B. (1987) Transcription of mouse rDNA is regulated by an activated subform of RNA polymerase I. *Cell*, **50**, 873.

Vamus,H.E. (1987) Oncogenes and transcriptional control. *Science*, **238**, 1337.

Vincent,A. (1986) TFIIIA and homologous genes. *Nucleic Acids Res.*, **14**, 4385.

Glossary

Activator: factor, usually a protein, binding at or near gene promoters and causing stimulation of transcription rate.

Allostery: modulation of enzyme activity involving conformational change of the protein after binding of the modulating agent (usually at a distinct 'allosteric site' on the protein surface).

Anti-termination: situation in which normal transcription termination at the end of a gene or operon is prevented by extraneous factors, allowing RNA synthesis to continue into other (downstream) genes.

Archaebacteria: third major grouping of the living (cellular) world, as distinct from the true bacteria (prokaryotes) and the eukaryotes.

Attenuation: transcriptional regulation acting to abort RNA synthesis after initiation but before read-through of protein-encoding sequences.

Box(es): a group of nucleotides that together form a sequence, usually a consensus sequence, which has a recognizable function.

Chromatin: complex of DNA and proteins, mainly histones, constituting the eukaryotic chromosomes.

***Cis*-acting:** modulators exerting their effects within the same molecule; examples are promoters and upstream elements which affect transcription rates of their cognate genes.

Closed complex: the complex of prokaryotic RNA polymerase and promoter DNA which forms first, prior to local unwinding of the DNA helix.

Consensus sequence: sequence of nucleotides compiled by comparison of homologous regions of many genes, and inserting the most frequently found one in each position.

Core enzyme: prokaryotic RNA polymerase with only β, β' and two α subunits but no σ subunit.

Core promoter: minimal sequence of promoter region needed to permit accurate initiation of transcription, as distinct from other (usually upstream) sequences which may modulate rates of initiation.

Discriminator sequence: GC-rich sequence found between Pribnow boxes and initiation sites of 'stable' (ribosomal and transfer) RNA genes in bacteria; thought to be involved in stringent control.

Enhancers: sequences of DNA found mainly in eukaryotes that can up-regulate transcription of neighbouring genes; however, they can exert their effects over long distances of intervening DNA, up or downstream of the gene and in either possible orientation.

Eukaryote: organism containing cells with distinct nuclei, mitochondria and other components distinctively different from the much simpler prokaryotes. Almost all organisms apart from bacteria and certain algae, including all differentiated multicellular ones, are eukaryotes.

Exon: that part of a eukaryotic gene coding for RNA which will ultimately be spliced into functional mRNA.

Finger protein: class of DNA-binding proteins characterized by loops ('fingers') of repeating amino acid sequences, each associated with a zinc atom, that bind in the major groove of the DNA helix.

Footprint: change observed in a DNA sequencing gel after the DNA has bound a specific protein, protecting some sequences from digestion by DNase I while sometimes enhancing the cleavage rates of others.

Fusion protein: protein derived from two genes artificially fused together and transcribed from a single promoter. Commonly produced by expression vectors after cloning and transformation into *E.coli*.

Gene: section of DNA which, in its entirety (including introns in eukaryotes), codes for a functional RNA molecule.

Genome: the total complement of DNA in an organism.

Half-life: time required for the concentration to fall to half the original value.

Half-time: time for a reaction to go to half completion.

Heat-shock genes: set of genes, usually few in number, which becomes active after sudden increases of temperature (usually of ~5°C). They are inactive at normal temperatures, but dominate the pattern of gene expression after the heat shock.

hnRNA: heterogeneous nuclear RNA; a complex mixture of large RNAs present in eukaryotic cell nuclei and including the precursors to functional species, especially mRNAs.

Hogness box: see TATA box.

Holoenzyme: the entire functional prokaryotic RNA polymerase, including σ subunit.

Housekeeping gene: a gene more or less universally required by all or most tissues of a multicellular organism, and which is usually expressed constitutively (i.e. all the time). Examples include those coding for enzymes of ubiquitous metabolic pathways, such as glycolysis.

Initiation site: the precise base pair in the DNA at which transcription starts, corresponding to the 5'-end of the RNA molecule synthesized.

Intron: region of eukaryotic gene coding for RNA later removed during splicing and thus not contributing to the final functional RNA product.

m (messenger) RNA: RNA molecules including regions which code for proteins and which are translated on the ribosomes.

Negative control: regulation of transcription in which factors are normally present to prevent RNA synthesis; activity only occurs after removal of such factors (repressors).

Non-transcribed spacer (NTS): long segments of DNA between tandemly-re-iterated genes, especially ribosomal genes of eukaryotes, on which no transcriptional activity is usually seen in electron microscope preparations.

Open complex: stable complex formed between prokaryotic RNA polymerase holoenzyme and promoter regions after local melting of the DNA helix, just prior to initiation.

Operator: region of DNA between Pribnow box and protein-coding regions of those bacterial operons subject to negative control; it is the sequence to which the repressor binds.

Operon: section of DNA in which two or more related genes lie adjacent to one another and are transcribed from a single promoter into polycistronic mRNA. Common in bacteria, rare or unknown in eukaryotes.

Palindrome: DNA sequence which reads the same in both directions, taking account of the two strands. A simple example is:

$$5'-AAAAAATTTTTT-3'$$
$$3'-TTTTTTAAAAAA-5'$$

Plasmids: circular double-stranded DNA molecules commonly found in bacteria additional to the main chromosome; widely used for cloning purposes.

Polyadenylation: addition of a series of AMP residues at the 3'-end of most eukaryotic mRNAs, by a distinct enzyme and subsequent to transcription, to form a poly(A) tail.

Polycistronic mRNA: mRNA which encodes more than one protein, usually found in bacteria as a result of the transcription of operons.

Positive control: situation in which transcription is increased by the addition of an activator, often a protein factor.

Pribnow box: AT-rich region, about 7 bp long, central to prokaryotic promoters at about position −10 from the initiation site.

Prokaryote: simple cellular life-form without distinct nucleus; includes bacteria and certain (blue-green) algae.

Promoter: region of DNA crucial to the accuracy and rate of transcription initiation. Usually, but not always, immediately upstream of the gene itself.

Pseudopromoter: region of DNA where transcription may start artifactually *in vitro*, but not used *in vivo*. Single-strand breaks in DNA can be strong pseudopromoters.

Re-iterated genes: genes present in more than one copy per haploid genome (e.g. ribosomal genes).

Repressor: protein which binds to parts of promoter DNA (operator region) to prevent transcription; well-characterized examples are known from bacteria.

r (ribosomal) RNAs: types of RNA found in functional ribosomes; there are three species in prokaryotes (5S, 16S and 23S, all derived from a common 30S precursor) and four species in eukaryotes. In mammals these are 5S, 5.8S, 18S and 28S.

snRNAs: small nuclear RNAs, a family of small (<11S) RNAs found in eukaryotic nuclei. Many are of unknown function, but some at least (notably U1) are involved in the processing of hnRNAs to functional mRNAs.

Splicing: the processing of hnRNA in eukaryotic nuclei to remove introns and ligate exons, and hence form functional mRNA. Some other types of RNA (e.g. certain tRNAs) are similarly processed.

Stringency: the tight linkage, especially in bacteria, between the rate of protein synthesis and the rates of stable rRNA and tRNA) RNA transcription.

Supercoiling: extra tension in DNA generated by cutting one strand, revolving it round the axis of the uncut one, and resealing it in an otherwise constrained DNA molecule (i.e. a linear one in which the ends are not free to rotate, or one which is circular). Many types of DNA (e.g. bacterial plasmids) are supercoiled *in vivo*.

S-values: anotations commonly given to macromolecules, especially RNAs, which reflect their rate of sedimentation through a sucrose gradient. S-values reflect a composite of both the size and the shape of a molecule, and are not simply proportional to size alone.

Tandem linkage: situation in which re-iterated genes are spaced along a DNA molecule one after another. The classical example is the (nucleolar) ribosomal gene set of eukaryotes.

TATA box: AT-rich sequence common to many eukaryotic promoters of the type used by RNA polymerase II, situated at about -25 bp from the initiation site in higher eukaryotes but further away in yeasts. Otherwise known as the Hogness box and analogous to the Pribnow box of prokaryotes.

Terminator: DNA sequence causing termination of transcription at the 3'-end of a gene.

***Trans*-acting:** action to up- or down-regulate transcription by factors not part of the gene in question; thus transcription factors (proteins) are *trans*-acting.

Transcription factors: molecules, usually proteins, which are necessary for accurate initiation of transcription or which affect the rate of transcription but are not integral parts of RNA polymerases.

t (transfer) RNA: a family (usually $>$ 50 types per cell) of small (4S) RNA molecules that serve adaptor functions, bringing amino acids to the site of protein synthesis on the ribosome.

Unique genes: genes present only once per haploid genome. Most structural (protein-encoding) genes of eukaryotes and all those of prokaryotes fall into this category.

Upstream element: DNA sequence upstream of the core promoter which affects the rate of transcription, usually (probably always) in a way mediated by specific DNA-binding proteins.

Western blotting: transfer of proteins from a gel after electrophoresis to a membrane for subsequent analysis.

Index

Abortive initiation, 9–10, 47
Acanthamoeba, 55
   encystment of, 55
Actin gene, 60
Activator proteins, 4, 19
Adenovirus, 36, 42, 46–47
Adenyl cyclase, 19
Alcohol dehydrogenase gene, 60
Allostery, 28
Amino acid starvation, 23–27
α-amanitin gene, 42, 44
α-amylase gene, 60
Antennapedia gene, 60
Anti-termination, 5, 11–15, 22, 30
Antithrombin III gene, 38
AP1, 59
Attenuation, 21–23, 56

Bacillus subtilis, 28–31
   sporulation of, 28–31
Bacteriophage,
   infection, 30–33
   λ phage, 11–15, 19–21, 30
   SP01 phage, 30–31
   SP6 phage, 33
   T3 phage, 31
   T4 phage, 31
   T7 phage, 8, 10–11, 31–33, 45

CAAT box, 38–39, 46, 58–59, 64
Calcitonin gene, 60
Caps,
   on mRNA, 47, 49
Catabolite activator protein (CAP), 19
Chromatin, 39, 45, 47, 59, 62
Chromosomes, 35, 61
Class I genes, 36
Class II genes, 36
Class III genes, 39
Closed complex, 7–8
Collagen gene, 38
Conalbumin gene, 36

Core enzyme, 5, 10
Cyclic AMP, 17, 19
   receptor protein (CRP), *see*
      Catabolite activator protein
Cytochrome *c* genes, 38, 56

Dihydrofolate reductase gene, 38
Discriminator sequences, 26
DNA,
   bending, 3, 47
   dyad symmetry, 4, 19
   nuclease-sensitivity, 42
   palindromes, *see* dyad symmetry
   supercoiling, 8
   Z-DNA, 21
DNA-binding proteins,
   finger proteins, 65–67
   initiation factors, 46–47, 55–64
   repressors, 18–21, 23
DNase-sensitive sites, 42, 59
Drosophila,
   chromatin, 40–42
   finger proteins, 66
   gene numbers, 35
   heat shock, 59
   P-element, 60
   ribosomal genes, 39
   RNA polymerase, 43
Duchenne muscular dystrophy
   gene, 35
Dunce gene, 60

Elongation, 9–10
Embryogenesis, 61–64
Encystment, 55
Enhancers, 36, 38, 59
Escherichia coli,
   bacteriophage infection of, 31
   finger proteins, 66
   genome size, 1, 35
   growth control, 23
   lexA protein, 56

promoters, 7, 36
RNA polymerase structure, 5, 43
Exons, 35, 49–50, 60–61, 64

Fibroin gene, 46
Finger proteins, 64–67
Footprint (DNA), 8, 56

GAL1, -4, -80, 56–57
Gal operon, 19
Gene,
  numbers, 35
Globin genes, 35–36, 38, 46, 49, 59
GLN4, 56
Guanosine tetraphosphate, 24–28, 53

HAP1, 55–56
Heat shock responses, 29, 38, 40, 59
Heparin, 10
Heterogeneous nuclear RNA, 42, 49, 54
His operon, 21, 23
Histones, 39, 47, 62, 64
  genes, 46, 49
HMRE mating locus, 38
HnRNA, see Heterogeneous nuclear RNA
Holoenzyme, 5
Housekeeping genes, 38
Hypoxanthine phosphoribosyltransferase gene, 38, 59

IF2, 28
ilv operon, 23
Immunoglobulin genes, 38, 60
Inducer, 17–18, 56
Initiation of transcription,
  abortive, 9, 47
  eukaryote, 45–47
  prokaryote, 8–10
Intron, 35, 38, 49–50, 60–61, 64

jun oncoprotein, 56

Klebsiella pneumoniae, 3, 19
  nitrogen fixation genes, 3, 19

Lac operon, 1, 8, 35, 67
Lariats, 49–50
Leu operon, 21
lexA protein, 56–57
Lymphocytes, 53–54

Messenger RNA,
  bacteriophage, 30–31
  eukaryotic, 39, 49, 62

polycistronic, 1, 21
prokaryotic, 1, 21, 24–28, 56
Mouse,
  finger proteins, 66
  nerve growth factor gene, 60
  ribosomal genes, 55
  transgenic, 59
mRNA, see Messenger RNA

Nematodes, 50
Nerve growth factor gene, 60
Nitrogen fixation genes, 3, 19
N-protein, 11–15
Nucleosome, 40–42, 47
NusA, 12–15
Nut box, 4, 15

Oogenesis, 54, 61–64
Open complex, 7–8, 10, 46
Operator, 18, 56
Operons, 1, 17–19, 35
Ovalbumin gene, 38, 46

Palindromes, 4
Pausing, 11, 22
P-element, 60
Phe operon, 21
Phosphoglycerate kinase gene, 38
Polyadenylation, 60, 64
Polycistronic mRNA, 1, 31, 35
Polyoma virus, 38
Pribnow box, 2
Processing,
  of RNA, 47, 49–50, 53
Promoters,
  alternative, 60
  bacteriophage, 30–32
  eukaryotic, 36–39, 45–46
  prokaryotic, 1–3, 18–21, 24–33, 55
Protamine genes, 46
Pseudopromoters, 45

Rapid-start complex, 46
Rat,
  finger proteins, 66
  tropomyosin gene, 60–61
RC2, 56
relA gene, 24
Repressors, 4, 18–21, 23
Rho factor, 4, 11–15, 22
Ribosomes,
  eukaryote RNA genes, 35–37, 39–40, 42, 47, 61–66
  prokaryote RNA genes, 1, 13, 23–28
  proteins, 17, 24

RNA, 1, 35–36, 50, 53
Rifampicin, 7, 9, 31, 45
RNA polymerases,
   chloroplast, 43–45
   eukaryote,
      pol I, 35–36, 42–43, 46
      pol II, 35–39, 42–43, 46, 64
      pol III, 35, 39–40, 42–43, 47–48, 62
   mitochondrial, 43–45
   phosphorylation, 43, 55
   prokaryote, 5–15, 19–24, 26, 28–33
   subunit structures, 5–7, 42–45
   viral, 31–33
Run-off assay, 45

Sarkosyl, 46
Self-splicing, 50
S-factors, 47
Sgs4 gene, 42, 60
Sigma subunit, 5–7, 10, 13–14, 28–29, 31
Silencers, 38
Small nuclear RNAs, 35, 38, 42, 46, 50, 67
SP1 protein, 58–59, 66–67
Splicing, 50
Starvation stringency protein, 24
Stem-loop structures, 4, 11, 21
Stringent response, 23, 53
Supercoiled DNA, 8
SV40, 38–39, 46, 58–59

T-antigen, 38, 58–59
TATA box, 36, 38–39, 46, 59, 64
Termination,
   eukaryote, 36, 38–39, 47
   prokaryote, 2, 11–15, 22
Tetrahymena,
   self-splicing RNA, 50
TFIIIA, 47–48, 62–66
TFIIIB, 47–48, 62
TFIIIC, 47–48, 62
Thr operon, 21, 23
Thymidine kinase gene, 58–59
Transfer RNA, 1, 21–28, 35
   attenuation, 21–23
   gene organization, 1, 35
   genes,
      promoters, 39–40, 47, 67
      transcription, 23–28, 42, 47
Tropomyosin gene, 60–61
tRNA, *see* Transfer RNA
Trp operon, 21–23

Trypanosomes, 50, 67
Tyrosine aminotransferase gene, 38

Upstream activator sequences, 38, 55, 59

Variable surface glycoprotein genes, 67

Wastage of RNA, 54

Xenopus,
   finger proteins, 66
   oogenesis, 54, 61–66
   ribosomal gene transcription, 36, 39, 46, 47, 61–66

Yeast,
   finger proteins, 66–67
   promoters, 36
   RNA polymerases, 42–44
   transcription factors, 55–57
   upstream elements, 38, 39, 55

Z-DNA, 21

GENE STRUCTURE AND TRANSCRIPTION

IN FOCUS

Titles published in the series:

*Complement

Enzyme Kinetics

Gene Structure and Transcription

Genetic Engineering

*Immune Recognition

*Lymphokines

Molecular Basis of Inherited Disease

Regulation of Enzyme Activity

*Published in association with the British Society for Immunology.

Series editors

David Rickwood
Department of Biology, University of Essex, Wivenhoe Park, Colchester, Essex CO4 3SQ, UK

David Male
Institute of Psychiatry, De Crespigny Park, Denmark Hill, London SE5 8AF, UK